Stochastic Processes

Stochastic Processes
An introduction

P. W. Jones

Mathematics Department, Keele University

and

P. Smith

Mathematics Department, Keele University

A member of the Hodder Headline Group
LONDON
Co-published in the United States of America by
Oxford University Press Inc., New York

First published in Great Britain in 2001 by
Arnold, a member of the Hodder Headline Group,
338 Euston Road, London NW1 3BH

http://www.arnoldpublishers.com

Co-published in the United States of America by
Oxford University Press Inc.,
198 Madison Avenue, New York, NY10016

British Library Cataloguing in Publication Data
A catalogue record for this book is available from the British Library

Library of Congress Cataloging-in-Publication Data
A catalog record for this book is available from the Library of Congress

ISBN 0 340 80654 0

1 2 3 4 5 6 7 8 9 10

Production Editor: James Rabson
Production Controller: Martin Kerans
Cover Design: Terry Griffiths

Typeset in 10/12 pt Times by HK Typesetting Ltd, London
Printed and bound in Malta by Gutenberg Press Ltd

What do you think about this book? Or any other Arnold title?
Please send your comments to feedback.arnold@hodder.co.uk

Contents

Series preface

Arnold Texts in Statistics is a series that is designed to provide an introductory account of key subject areas in statistics. Each book focuses on a particular area, and subjects covered include statistical modelling, time series, applied stochastic modelling and randomized controlled clinical trials. Texts in this series combine theoretical development with practical examples. Indeed, a distinguishing feature of the texts is that they are copiously illustrated with examples drawn from a range of applications. These illustrations take full account of the widespread availability of statistical packages and other software for data analysis. The theoretical content of the texts is sufficient for an appreciation of the techniques being presented, but mathematical detail is included only when necessary. The texts are designed to be accessible to undergraduate and postgraduate students of statistics. In addition, they will enable statisticians and quantitative scientists working in research institutes, industry, government organizations, market research agencies, financial institutions, and so on, to update their knowledge in those areas of statistics that are of direct relevance to them.

David Collett
Series Editor

Preface

This textbook has developed from a course in stochastic processes given by the authors over the past 20 years to second-year students studying Mathematics or Statistics at Keele University, UK. At Keele, the great majority of students are taking joint degrees in Mathematics or Statistics and another subject, which may be from the sciences, social sciences or the humanities. For this reason, the course has always had to appeal to students of varied academic interests, and this is reflected in this book by applications and examples that students can quickly understand and relate to. Specialized applications have been avoided to accord with our view that students have enough to contend with in the mathematics required in stochastic processes.

Topics can be selected from Chapters 2 to 9 for a one-semester course or module in stochastic processes. It is assumed that readers have already encountered the usual first-year courses in calculus and matrix algebra, and also a first course in probability; however, a revision of relevant basic probability is included for reference in Chapter 1. Some of the easier material on discrete random processes is included in Chapters 2, 3 and 4, which cover some simple gambling problems, random walks and population models. Random processes continuous in time are developed in Chapters 5 and 6. These include Poisson processes and population models. Chapter 7 presents an introduction to queues viewed as stationary processes. The book ends with two chapters on reliability and other random processes, the latter including branching processes, martingales and a simple epidemic.

There are over 250 worked examples and problems for the reader to attempt, and answers and hints to selected problems are included in the Appendix. Mathematical software such as *Mathematica* has become an integral part of most degree courses in Mathematics or Statistics. For this reason, several projects are presented in Chapter 10. They are intended to illustrated the text both as mathematical and numerical examples. Possible programs in *Mathematica* for each of the projects can be downloaded from the Keele University, Mathematics Department website at:

http://www.keele.ac.uk/depts/ma/

Not every topic in the book is included, but the programs, which generally use standard *Mathematica* commands, are intended to be flexible in that inputs, parameters, data, etc. can be varied. Graphs and computations can often add insight into what might otherwise be viewed as rather mechanical analyses. In addition, more complicated examples, which are really beyond hand calculations, can be attempted.

Finally, we would like to thank the many students at Keele over many years who have helped to develop this course.

Peter Jones
Peter Smith

April 2001

1
Some background on probability

1.1 Introduction

We shall be concerned with the modelling of random experiments using the theory of probability. This is known generally as the study of *stochastic* or *random processes*. In particular, we shall be interested in the way in which the results or outcomes of these experiments vary or evolve over time. An *experiment* or *trial* is any situation where an outcome is observed. In many of the applications considered, these outcomes will be numerical, sometimes in the form of counts. The experiment is random if the outcome is not predictable or is uncertain.

First, we are going to be concerned with simple mechanisms for creating random outcomes, namely games of chance. Initially, one recurring theme will be the study of the classical problem known as *gambler's ruin*. We will then move on to applications of probability to modelling in, for example, engineering, medicine and biology. We make the assumption that the reader is familiar with the basic theory of probability. This background will, however, be reinforced by the brief review of these concepts, which will form the main part of this chapter.

1.2 Probability

In random experiments, the list of all possible outcomes is termed the *sample space*, S. This consists of individual *outcomes* or *elements*. These elements have the properties that they are *mutually exclusive* and that they are *exhaustive*. Mutually exclusive means that two or more outcomes cannot occur simultaneously: exhaustive means that all possible outcomes are in the list. Thus, each time the experiment is carried out one of the outcomes in S must occur. A collection of elements of S is called an *event*: these are usually denoted by capital letters, A, B, etc. We denote by $\mathbf{P}(A)$ the *probability* that the event A will occur at each repetition of the random experiment. Remember that A is said to have occurred if one element making up A has occurred. In order to calculate or estimate the probability of an event A there are two possibilities. In one approach an experiment can be performed a large number of times, and $\mathbf{P}(A)$ can be approximated by the relative frequency with which A occurs. In order to analyse random experiments we make the assumption that the conditions surrounding the trials remain the same. We hope that some regularity or settling down of the outcome is apparent. The ratio

$$\frac{\text{the number of times a particular outcome } A \text{ occurs}}{\text{total number of trials}}$$

is known as the *relative frequency* of the outcome, and the number to which it appears to converge as the number of trials increases is known as the probability of the outcome of A. Where we have a finite sample space it might be reasonable to assume that the outcomes of an experiment are equally likely to occur as in the case, for example, of rolling a fair die or spinning an unbiased coin. In this case the probability is calculated by counting in S giving for A,

$$\mathbf{P}(A) = \frac{\text{number of elements of } S \text{ where } A \text{ occurs}}{\text{number of elements in } S}.$$

There are, of course, many 'experiments' which are not repeatable. Horse races are only run once, and the probability of a particular horse winning a particular race may not be calculated by relative frequency. However, a punter may form a view about the horse based on other factors that may be repeated over a series of races. The past form of the horse, the form of other horses in the race, the state of the course, the record of the jockey, etc, may all be taken into account in determining the probability of a win. This leads to a view of probability as a 'degree of belief' about uncertain outcomes. The odds placed by bookmakers on the horses in a race reflect how punters place their bets on the race.

It is usually convenient to use *set* notation when deriving probabilities of events. This leads to S being termed the *universal set*: A is called a *subset* of S. This also helps with the construction of more complex events in terms of the unions and intersections of several events (that is, sets). The Venn diagrams shown in Figure 1.1 show the main set operations of *union* ($\cup$), *intersection* ($\cap$) and *complement* (A^c) which are required in probability.

- **Union**. The union of two sets A and B is the set of all elements that belong to A, or to B, or to both. It can be written formally as

$$A \cup B = \{x | x \in A \text{ or } x \in B \text{ or both}\}.$$

- **Intersection**. The intersection of two sets A and B is the set $A \cap B$ that contains all elements common to both A and B. It can be written as

$$A \cap B = \{x | x \in A \text{ and } x \in B\}.$$

- **Complement**. The complement A^c of a set A is the set of all elements that belong to the universal set S but do not belong to A. It can be written as

$$A^c = \{x \notin A\}.$$

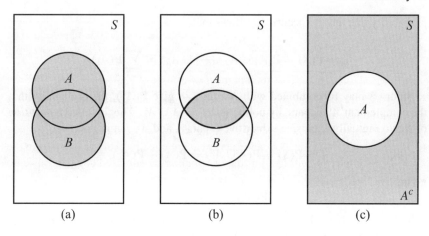

Fig. 1.1 *(a) The union $A \cup B$ of A and B; (b) the intersection $A \cap B$ of A and B; (c) the complement A^c of A: S is the universal set.*

So, for example, in an experiment in which we are interested in two events A and B, then $A^c \cap B$ may be interpreted as 'only B', being the intersection of the *complement* of A and B. How does that appear on a Venn diagram? We denote by $\varnothing$ the *empty set*, that is the set which contains no elements. Note that $S^c = \varnothing$. Two events A and B are said to be *mutually exclusive* (that is, A and B are *disjoint sets* in set terminology) if $A \cap B = \varnothing$.

The probability of any event satisfies the three axioms

- **Axiom 1**: $0 \leq \mathbf{P}(A) \leq 1$ for every event A
- **Axiom 2**: $\mathbf{P}(S) = 1$
- **Axiom 3**: $\mathbf{P}(A \cup B) = \mathbf{P}(A) + \mathbf{P}(B)$ if A and B are mutually exclusive $(A \cap B = \varnothing)$

Axiom 3 may be extended to more than two mutually exclusive events, say k of them represented by

$$A_1, A_2, \ldots, A_k$$

where $A_i \cap A_j = \varnothing$ for all $i \neq j$. This is called a *partition* of S if

(a) $A_i \cap A_j = \varnothing$ for all $i \neq j$,

(b) $\bigcup_{i=1}^{k} A_i = A_1 \cup A_2 \cup \ldots \cup A_k = S$,

(c) $\mathbf{P}(A_i) > 0$.

In this definition, (a) states that the events are mutually exclusive, (b) that every event in S occurs in one of the events A_i, and (c) implies that there is a non-zero probability that any A_i occurs. It follows that

$$1 = \mathbf{P}(S) = \mathbf{P}(A_1 \cup A_2 \cup \cdots \cup A_k) = \sum_{i=1}^{k} \mathbf{P}(A_i).$$

Axiom 3 may be combined with Axiom 2 to give $\mathbf{P}(A^c)$, the probability that the complement A^c occurs, by noting that $S = A \cup A^c$. This is called a *partition* of S into mutually exclusive exhaustive events A and A^c. Thus

$$1 = \mathbf{P}(S) = \mathbf{P}(A \cup A^c) = \mathbf{P}(A) + \mathbf{P}(A^c),$$

giving
$$\mathbf{P}(A^c) = 1 - \mathbf{P}(A).$$

In the case where A and B are not mutually exclusive (that is, $A \cap B \neq \varnothing$), then it may be proved, by using the Axioms, that

$$\mathbf{P}(A \cup B) = \mathbf{P}(A) + \mathbf{P}(B) - \mathbf{P}(A \cap B). \tag{1.1}$$

Example 1.1. Two distinguishable fair dice a and b are rolled and the values on the uppermost faces noted. What are the elements of the sample space? What is the probability that the sum of the face values of the two dice is 7? What is the probability that at least one 5 appears?

We distinguish first the outcome of each die so that there are $6 \times 6 = 36$ possible outcomes for the pair. The sample space has 36 elements of the form (i, j) where i and j take all integer values $1, 2, 3, 4, 5, 6$, and i is the outcome of die a and j is the outcome of b. The full list is

$$\begin{aligned}
S = \{&(1, 1), (1, 2), (1, 3), (1, 4), (1, 5), (1, 6),\\
&(2, 1), (2, 2), (2, 3), (2, 4), (2, 5), (2, 6),\\
&(3, 1), (3, 2), (3, 3), (3, 4), (3, 5), (3, 6),\\
&(4, 1), (4, 2), (4, 3), (4, 4), (4, 5), (4, 6),\\
&(5, 1), (5, 2), (5, 3), (5, 4), (5, 5), (5, 6),\\
&(6, 1), (6, 2), (6, 3), (6, 4), (6, 5), (6, 6)\},
\end{aligned}$$

and they are all assumed to be equally likely since the dice are fair. If A_1 is the event that the sum of the dice is 7, then from the table,

$$A_1 = \{(1, 6), (2, 5), (3, 4), (4, 3), (5, 2), (6, 1)\}$$

which occurs for six elements out of 36. Hence

$$P(A_1) = \tfrac{6}{36} = \tfrac{1}{6}.$$

The event that at least one 5 appears is the list

$$A_2 = \{(1, 5), (2, 5), (3, 5), (4, 5), (5, 1), (5, 2), (5, 3), (5, 4), (5, 5), (5, 6), (6, 5)\},$$

which has 11 elements. Hence

$$P(A_2) = \frac{11}{36}.$$

$\square$

Example 1.2. From a well-shuffled pack of 52 playing cards a single card is randomly drawn. Find the probability that it is a heart or an ace.

Let A be the event that the card is an ace, and B the event that it is a heart. The event $A \cap B$ is the ace of hearts. We require the probability that it is an ace or a heart, which is $\mathbf{P}(A \cup B)$. However, since one of the aces is a heart the events are not mutually exclusive. Hence, we must use equation (1.1). It follows that

the probability that an ace is drawn is $\mathbf{P}(A) = 4/52$,

the probability that a heart is drawn is $\mathbf{P}(B) = 13/52 = 1/4$,

the probability that the ace of hearts is drawn is $\mathbf{P}(A \cap B) = 1/52$.

From (1.1)

$$\mathbf{P}(A \cup B) = \mathbf{P}(A) + \mathbf{P}(B) - \mathbf{P}(A \cap B) = \frac{4}{52} + \frac{1}{4} - \frac{1}{52} = \frac{16}{52} = \frac{4}{13}.$$

This example illustrates events that are not mutually exclusive. The result could also be obtained directly by noting that 16 of the 52 cards are either hearts or aces.

In passing, note that $A \cap B^c$ is the set of aces excluding the ace of hearts, whilst $A^c \cap B$ is the heart suit excluding the ace of hearts. Hence

$$\mathbf{P}(A \cap B^c) = 3/52, \qquad \mathbf{P}(A^c \cap B) = 12/52.$$

$\square$

1.3   Conditional probability and independence

If the occurrence of an event B is affected by the occurrence of another event A then we say that A and B are dependent events. We might be interested in a random experiment with which A and B are associated. When the experiment is performed, it is known that event A has occurred. Does this affect the probability of B? Generally the answer is yes: there may be the dependence between this probability and the events. This probability of B now becomes the *conditional probability* of B given A, which is now written as $\mathbf{P}(B|A)$. Usually this will be distinct from the probability $\mathbf{P}(B)$. Strictly speaking, this probability is conditional since we must assume that B is conditional on the sample space occurring, but it is implicit in $\mathbf{P}(B)$. On the other hand, the conditional probability of B is restricted to that part of the sample space where A has occurred. This conditional probability is defined as

$$\mathbf{P}(B|A) = \frac{\mathbf{P}(A \cap B)}{\mathbf{P}(A)}, \qquad \mathbf{P}(A) \neq 0. \tag{1.2}$$

If the probability of B is unaffected by the prior occurrence of A then we say that A and B are *independent* or that

$$\mathbf{P}(B|A) = \mathbf{P}(B),$$

which, from the above, implies that

$$\mathbf{P}(A \cap B) = \mathbf{P}(A)\mathbf{P}(B).$$

Conversely, if $\mathbf{P}(B|A) = \mathbf{P}(B)$, then A and B are independent events. Again this result can be extended to three or more independent events.

Example 1.3. Let A and B be independent events with $\mathbf{P}(A) = \frac{1}{4}$ and $\mathbf{P}(B) = \frac{2}{3}$. Calculate the following probabilities:

  (a) $\mathbf{P}(A \cap B)$;
  (b) $\mathbf{P}(A \cap B^c)$;
  (c) $\mathbf{P}(A^c \cap B^c)$;
  (d) $\mathbf{P}(A^c \cap B)$;
  (e) $\mathbf{P}((A \cup B)^c)$.

Since the events are independent, then $\mathbf{P}(A \cap B) = \mathbf{P}(A)\mathbf{P}(B)$. Hence

  (a) $\mathbf{P}(A \cap B) = \frac{1}{4} \cdot \frac{2}{3} = \frac{1}{6}$.
  (b) We can express the event A as

$$A = (A \cap B) \cup (A \cap B^c).$$

It follows that

$$\mathbf{P}(A \cap B^c) = \mathbf{P}(A) - \mathbf{P}(A \cap B) = \mathbf{P}(A) - \mathbf{P}(A)\mathbf{P}(B),$$

since A and B are independent events. Hence

$$\mathbf{P}(A \cap B^c) = \mathbf{P}(A)[1 - \mathbf{P}(B)] = \mathbf{P}(A)\mathbf{P}(B^c).$$

Inserting the given probabilities we find that

$$\mathbf{P}(A \cap B^c) = \mathbf{P}(A)\mathbf{P}(B^c) = \mathbf{P}(A)[1 - \mathbf{P}(B)] = \frac{1}{4}(1 - \frac{2}{3}) = \frac{1}{12}.$$

  (c) Since A^c and B^c are independent events,

$$\mathbf{P}(A^c \cap B^c) = \mathbf{P}(A^c)\mathbf{P}(B^c) = [1 - \mathbf{P}(A)][1 - \mathbf{P}(B)] = \frac{3}{4} \cdot \frac{1}{3} = \frac{1}{4}.$$

(d) $\mathbf{P}(A^c \cap B) = \mathbf{P}(A^c)\mathbf{P}(B) = [1 - \frac{1}{4}]\frac{2}{3} = \frac{1}{2}.$

(e) $\mathbf{P}((A \cup B)^c) = 1 - \mathbf{P}(A \cup B) = 1 - \mathbf{P}(A) - \mathbf{P}(B) + \mathbf{P}(A \cap B)$ by equation (1.1). Hence

$$\mathbf{P}((A \cup B)^c) = 1 - \mathbf{P}(A) - \mathbf{P}(B) + \mathbf{P}(A)\mathbf{P}(B) = 1 - \frac{1}{4} - \frac{2}{3} + \frac{1}{6} = \frac{1}{4}.$$

$\square$

Example 1.4. For three events A, B and C, show that

$$\mathbf{P}(A \cap B | C) = \mathbf{P}(A | B \cap C)\mathbf{P}(B | C).$$

By using (1.2) and viewing $A \cap B \cap C$ as $(A \cap B) \cap C$ or $A \cap (B \cap C)$,

$$\mathbf{P}(A \cap B \cap C) = \mathbf{P}(A \cap B | C)\mathbf{P}(C) = \mathbf{P}(A | B \cap C)\mathbf{P}(B \cap C).$$

Hence

$$\mathbf{P}(A \cap B | C) = \mathbf{P}(A | B \cap C)\frac{\mathbf{P}(B \cap C)}{\mathbf{P}(C)} = \mathbf{P}(A | B \cap C)\mathbf{P}(B | C)$$

by (1.2) again.

$\square$

A result known as the *law of total probability*, or the *partition theorem*, will be used extensively later, for example, in the discrete gambler's ruin problem (Section 2.1) and the Poisson process (Section 5.2). Suppose that $A_1, A_2, \ldots, A_k$ represents a partition of S into k mutually exclusive events in which, interpreted as sets, the sets fill the space S but with none of the sets overlapping. Figure 1.2 shows such a scheme. When a random experiment takes place one and only one of the events can take place.

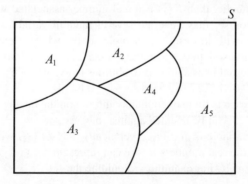

Fig. 1.2 *Schematic set view of a partition of S into 5 events $A_1, \ldots, A_5$.*

Suppose that B is another event associated with the same random experiment (Figure 1.2). Then B must be made up of the sum of the intersections of B with each of the events in the partition. Some of these will be empty but this does not matter. We can say that B is the union of the intersections of B with each A_i. Thus

$$B = \bigcup_{i=1}^{k} B \cap A_i,$$

but the significant point is that any pair of these events is mutually exclusive. By equation (1.1), it follows that

$$\mathbf{P}(B) = \sum_{i=1}^{k} \mathbf{P}(B \cap A_i). \tag{1.3}$$

Since, from equation (1.2),

$$\mathbf{P}(B \cap A_i) = \mathbf{P}(B|A_i)\mathbf{P}(A_i),$$

equation (1.3) can be expressed as

$$\mathbf{P}(B) = \sum_{i=1}^{k} \mathbf{P}(B|A_i)\mathbf{P}(A_i),$$

which is the law of total probability or the partition theorem.

1.4 Discrete random variables

In most of the applications considered in this text, the outcome of the experiment will be numerical. A *random variable* – usually denoted by the capital letters X, Y or Z, say – is a numerical value associated with the outcome of a random experiment. If s is an element of the original sample space S, which may be numerical or symbolic, then $X(s)$ is a real number associated with s. The same experiment, of course, may generate several random variables. Each of these random variables will, in turn, have sample spaces whose elements are usually denoted by lower case letters, such as $x_1, x_2, x_3, \ldots$ for the random variable X. We are now interested in assigning probabilities to events such as $\mathbf{P}(X = x_2)$ and $\mathbf{P}(X \leq x_3)$

If the sample space is finite or countably infinite on the integers (that is, the elements $x_0, x_1, x_2, \ldots$ can be counted against integers, say 0, 1, 2, ...) then we say that the random variable is *discrete*. Technically, the set $\{x_i\}$ will be a countable subset $\mathbb{V}$, say, of the real numbers $\mathbb{R}$. We can represent the $\{x_i\}$ generically by the variable x with $x \in \mathbb{V}$. For example, $\mathbb{V}$ could be the set

$$0, \tfrac{1}{2}, 1, \tfrac{3}{2}, 2, \tfrac{5}{2}, 3, \ldots.$$

In many cases $\mathbb{V}$ consists simply of the integers or a subset of the integers, such as

$$\mathbb{V} = \{0, 1\} \quad \text{or} \quad \mathbb{V} = \{0, 1, 2, 3, \ldots\}.$$

In the random walks of Chapter 3, however, $\mathbb{V}$ may contain all the positive and negative integers

$$\ldots - 3, -2, -1, 0, 1, 2, 3, \ldots.$$

In these *integer* cases we can put $x_i = i$.

As with most mathematical subjects, there is no universal notation in use: notation is often adapted to what seems to be the most useful and concise in the particular application.

We define

$$p(x_i) = \mathbf{P}(X = x_i), \text{ or, alternatively } p(x) \text{ or } p_x = \mathbf{P}(X = x),$$

which is known as the *probability mass function*. The pairs $\{x_i, p(x_i)\}$ for all i in the sample space define the *probability distribution* of the random variable X. If $x_i = i$, which occurs frequently in applications, then $p(x_i) = p(i)$ is replaced by p_i. Since the x values are mutually exhaustive then it follows from the axioms that

(a) $0 \leq p(x_i) \leq 1$ for all i,

(b) $\displaystyle\sum_{i=0}^{\infty} p(x_i) = 1$, or in generic form $\displaystyle\sum_{x \in \mathbb{V}} p(x) = 1$,

(c) $\mathbf{P}(X \leq x_k) = \displaystyle\sum_{i=0}^{k} p(x_i)$.

Example 1.5. A fair die is rolled until the first 6 appears face up. Find the probability that the first 6 appears at the xth throw.

Let the random variable X be the number of throws until the first 6 appears face up. This is an example of a discrete random variable X with an infinite number of possible outcomes

$$\{1, 2, 3, \ldots\}.$$

The probability of a 6 appearing for any throw is $\frac{1}{6}$ and of any other number appearing is $\frac{5}{6}$. Hence the probability of $x - 1$ numbers other than 6 appearing followed by a 6 is

$$\mathbf{P}(X = x) = \left(\frac{5}{6}\right)^{x-1}\left(\frac{1}{6}\right) = \frac{5^{x-1}}{6^x}.$$

$\square$

In the random processes considered in this text, it is sometimes necessary to indictate that the mass function is varying with time t. Notations such as $p_t(x_i)$ or $p(x_i, t)$ are used to emphasize that the random process is changing over time. If the context is obvious, the t is sometimes dropped.

1.5 Continuous random variables

In many applications, the discrete random variable, which for example might take the integer values $1, 2, \ldots,$ is inappropriate for problems where the random variable can take any real value in an interval. For example, the random variable T could be the time measured from time $t = 0$ until a light bulb fails. This could be any value $t \geq 0$. In this case, T is called a *continuous random variable*. Generally, if X is a continuous random variable there are mathematical difficulties in defining the event $X = x$: the probability is usually defined to be zero. Probabilities for continuous random variables may only be defined over intervals of values as, for example, in $\mathbf{P}(x_1 < X < x_2)$.

We define a *probability density function* (pdf) as $f(x)$ over $-\infty < x < \infty,$ which has the properties:

(a) $f(x) \geq 0, (-\infty < x < \infty);$

(b) $\mathbf{P}(x_1 < X < x_2) = \displaystyle\int_{x_1}^{x_2} f(x)\, dx$ for any x_1, x_2 such that $-\infty < x_1 < x_2 < \infty;$

(c) $\displaystyle\int_{-\infty}^{\infty} f(x)\, dx = 1.$

A typical graph of a density function $f(x)$ is shown in Figure 1.3. By (a) above the curve must remain non-negative, by (b) the probability that X lies between x_1 and x_2 is the shaded area, and by (c) the total area under the curve must be 1 since $\mathbf{P}(-\infty < X < \infty) = 1.$

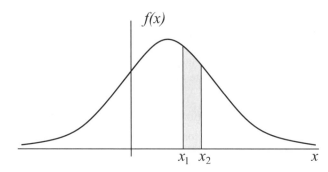

Fig. 1.3 *A probability density function.*

We can define the *cumulative distribution function* (cdf) $F(x)$ as the probability that X is less than or equal to x. Thus

$$F(x) = \mathbf{P}(X \leq x) = \int_{-\infty}^{x} f(u)\, du.$$

It follows from (c) above that

$$F(x) \to 1 \quad \text{as} \quad x \to \infty,$$

and that

$$\mathbf{P}(x_1 \leq X \leq x_2) = \int_{x_1}^{x_2} f(x)\,\mathrm{d}x = F(x_2) - F(x_1).$$

A typical cdf is shown in Figure 1.4.

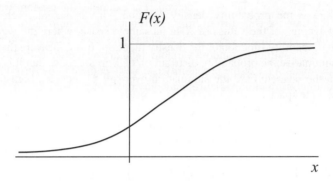

Fig. 1.4 *A cumulative distribution function.*

Example 1.6. Show that

$$f(x) = \begin{cases} 1/(b-a) & a \leq x \leq b \\ 0 & \text{for all other values of } x \end{cases}$$

is a possible probability density function. Find its cumulative distribution function.

The function $f(x)$ must satisfy conditions (a) and (c) above. This is the case since $f(x) \geq 0$ and

$$\int_{-\infty}^{\infty} f(x)\,\mathrm{d}x = \int_{a}^{b} \frac{1}{b-a}\,\mathrm{d}x = 1.$$

In addition, its cumulative distribution function $F(x)$ is given by

$$F(x) = \int_{a}^{x} \frac{1}{b-a}\,\mathrm{d}x = \frac{x-a}{b-a} \quad \text{for} \quad a \leq x \leq b.$$

For $x < a$, $F(x) = 0$ and for $x > b$, $F(x) = 1$.

The pdf $f(x)$ is the density function of the *uniform distribution*. □

1.6 Mean and variance

The *mean* (or *expectation* or *expected value*), $\mathbf{E}(X)$, of a random variable X is defined as

$$\mu = \mathbf{E}(X) = \sum_{i=0}^{\infty} x_i\, p(x_i) \text{ or } \sum_{x \in \mathbb{V}} x p(x), \text{ if } X \text{ is discrete,}$$

where $p(x_i) = \mathbf{P}(X = x_i)$ or $p(x) = \mathbf{P}(X = x)$, and

$$\mu = \mathbf{E}(X) = \int_{-\infty}^{\infty} xf(x)\,dx \text{ if } X \text{ is continuous,}$$

where $f(x)$ is the probability density function. It can be interpreted as the weighted average of the values of X in its sample space, where the weights are either the probability function or the density function. It is a measure that may be used to summarize the probability distribution of X in the sense that it is a central value. In the discrete case the summation over 'all x' includes both finite and infinite sample spaces.

A measure that is used in addition to the mean as a summary measure is the *variance* of X denoted by $\mathbf{V}(X)$. This gives a measure of variation or spread (dispersion) of the probability distribution of X, and is defined by

$$\sigma^2 = \mathbf{V}(X) = \mathbf{E}[(X - \mathbf{E}(X))^2] = \mathbf{E}[(X - \mu)^2]$$

$$= \begin{cases} \sum_{i=0}^{\infty}(x_i - \mu)^2 p(x_i) \text{ or } \sum_{x \in \mathbb{V}}(x - \mu)^2 p(x), & \text{if } X \text{ is discrete,} \\ \int_{-\infty}^{\infty}(x - \mu)^2 f(x)\,dx, & \text{if } X \text{ is continuous} \end{cases}$$

The variance is the mean of the squared deviations of each value of X from the central value μ. In order to give a measure of variation that is in the same units as the mean, the square root σ of $\mathbf{V}(X)$ is used. This is known as the *standard deviation* (sd) of X.

A function of a random variable is itself a random variable. If $h(X)$ is a function of the random variable X, then it can be shown that the expectation of $h(X)$ is given by

$$\mathbf{E}[h(X)] = \begin{cases} \sum_{i=0}^{\infty} h(x_i)p(x_i) \text{ or } \sum_{x \in \mathbb{V}} h(x)p(x), & \text{if } X \text{ is discrete} \\ \int_{-\infty}^{\infty} h(x)f(x)\,dx, & \text{if } X \text{ is continuous} \end{cases}$$

It is relatively straightforward to derive the following results for the expectation and variance of a linear function of X:

$$\mathbf{E}(aX + b) = a\mathbf{E}(X) + b = a\mu + b,$$

$$\mathbf{V}(aX+b) = \mathbf{E}[(aX+b-a\mu-b)^2] = \mathbf{E}[(aX-a\mu)^2] = a^2\mathbf{E}[(X-\mu)^2] = a^2\mathbf{V}(X),$$

where a and b are constants. Note that the *translation* of the random variable does not affect the variance. Also

$$\mathbf{V}(X) = \mathbf{E}(X^2) - \mu^2,$$

which is sometimes known as the *computational formula for the variance*. These results hold whether X is discrete or continuous and enable us to interpret expectation and variance as operators on X. Of course, in all cases, mean and variance only exist where the summations and infinite integrals are finite.

For expectations, it can be shown more generally that

$$\mathbf{E}\left[\sum_{i=1}^{k} a_i h_i(X)\right] = \sum_{i=1}^{k} a_i \mathbf{E}[h_i(X)],$$

where a_i, $i = 1, 2, \ldots, k$ are constants and $h_i(X)$, $i = 1, 2, \ldots, k$ are functions of the random variable X.

1.7 Some standard discrete probability distributions

In this section, we shall look at discrete random variables X with probability mass function $p(x)$ where x takes integer values. Each of the random variables considered are numerical outcomes of independent repetitions (or trials) of a simple experiment knowns as a *Bernoulli experiment*. This is an experiment where there are only two outcomes: a 'success' ($X = 1$) or a 'failure' ($X = 0$). The value of the random variable X is used as an *indicator* of the outcome, which may also be interpreted as the presence or absence of a particular characteristic. For example, in a single coin toss $X = 1$ is associated with the occurrence, or the presence of the characteristic, of a *head* and $X = 0$ with a *tail*, or the absence of a head.

The probability function of this random variable may be expressed in algebraic terms as

$$p_x = p^x q^{1-x}, \text{ over the two values } x = 0, 1,$$

where $p = \mathbf{P}(X = 1)$ is known as the *parameter* of the probability distribution, and $q = 1 - p$. We say that the random variable has a *Bernoulli distribution*.

The expected value of X is easily seen to be

$$\mu = \mathbf{E}(X) = 0 \times q + 1 \times p = p,$$

and the variance of X is

$$\sigma^2 = \mathbf{V}(X) = \mathbf{E}(X^2) - \mu^2 = 0^2 \times q + 1^2 \times p - p^2 = pq.$$

Suppose now that we are interested in random variables associated with independent repetitions of Bernoulli experiments, each with a probability of success, p. Consider first the probability distribution of a random variable X (a different X from that defined above), which is the number of successes in a fixed number of independent trials, n. Independence of the trials gives the probability of any sequence of 1s and 0s without order, which are mutually exclusive, leading to

$X = x$ as $p^x q^{n-x}$. The number of ways in which x successes can be arranged in n trials is

$$\frac{n!}{x!(n-x)!} = \binom{n}{x}.$$

Since each of these mutually exclusive sequences occurs with probability $p^x q^{n-x}$, the probability function of this random variable is given by

$$p(x) \text{ or } p_x = \binom{n}{x} p^x q^{n-x}, \qquad x = 0, 1, 2, \ldots, n. \tag{1.4}$$

This is the xth term in the binomial expansion of $(p+q)^n$, and for this reason is known as the *binomial distribution* with parameters n and p. A consequence of this observation is that

$$\sum_{x=0}^{n} p(x) = \sum_{x=1}^{n} \binom{n}{x} p^x q^{n-x} = (p+q)^n = 1.$$

The mean and variance may be easily shown to be np and npq respectively, which is n times the mean and variance of the Bernoulli distribution.

Instead of fixing the number of trials, suppose now that the number of successes, r, is fixed, and that the sample size required in order to reach this fixed number is the random variable X: this is sometimes called *inverse sampling*. Then, in the case $r = 1$, using the independence argument again, this leads to

$$p_x = q^{x-1} p, \qquad x = 1, 2, \ldots.$$

which is the *geometric probability function* with parameter p. Note that successive probabilities form a geometric series with common ratio $q = 1 - p$. Note that the sample space is now countably infinite. After some algebra it can be shown that the mean is given by $\mu = 1/p$, and the variance by $\sigma^2 = q/p^2$.

The geometric probability distribution possesses an interesting property known as the 'no memory' property, which can be expressed by

$$\mathbf{P}(X > a + b | X > a) = \mathbf{P}(X > b),$$

where a and b are positive integers. What this means is that if a particular event has *not* occurred in the first a repetitions of the experiment, then the probability that it will occur in the next b repetitions is the same as in the first b repetitions of the experiment. The result can be proved as follows. Using the definition of conditional probabilty in Section 1.3

$$\mathbf{P}(X > a + b | X > a) = \frac{\mathbf{P}(X > a + b \cap X > a)}{\mathbf{P}(X > a)} = \frac{\mathbf{P}(X > a + b)}{\mathbf{P}(X > a)}.$$

Since $\mathbf{P}(X > x) = q^x$,

$$\mathbf{P}(X > a + b | X > a) = \frac{q^{a+b}}{q^a} = q^b = \mathbf{P}(X > b).$$

The converse is also true, but the proof is not given here.

In the case where $r > 1$, the probability function of the number of trials may be derived by noting that $X = x$ requires that the xth trial results in the rth success and that the remaining $r - 1$ successes may occur in any order in the remaining $x - 1$ trials. Arguments based on counting the number of possible sequences of 1s and 0s, and the independence of trials, lead to

$$p(x) = \binom{x-1}{r-1} p^x q^{x-r}, \qquad x = r, r+1, \ldots.$$

This is known as a *Pascal* or *negative binomial distribution*. Its mean is r/p and its variance is rq/p^2, which are, respectively, r times the mean and the variance of the geometric distribution. Hence, a similar relationship exists between the geometric and the Pascal distributions as between the Bernoulli and the binomial distributions.

The binomial random variable arises as the result of observing n independent identically distributed Bernoulli random variables, and the Pascal by observing r sets of geometric random variables.

Certain problems involve the counting of the number of events that have occurred in a fixed time period. For example, the number of emissions of alpha particles by an X-ray source or the number of arrivals of customers joining a queue. It has been found that the *Poisson distribution* is appropriate in modelling these counts when the underlying process generating them is considered to be completely random. We shall spend some time in Chapter 5 defining such a process, which is known as a *Poisson process*.

As well as being a probability distribution in its own right, the Poisson distribution also provides a convenient approximation to the binomial distribution, to which it converges when n is large and p is small and $np = \alpha$, a constant. This is a situation where rounding errors would be likely to cause computational problems.

The Poisson probability function with parameter α is

$$p(x) = \frac{e^{-\alpha} \alpha^x}{x!}, \qquad x = 0, 1, 2, \ldots$$

with mean and variance both equal to α.

The *discrete uniform distribution* with integer parameter n has a random variable X that can take the values $x = r, r+1, r+2, \ldots r+n-1$ with the same probability $1/n$ (the continuous uniform distribution was introduced in Example 1.6). It is easy to show that the mean and variance of X are given by

$$\mu = \tfrac{1}{2}(n+1), \qquad \sigma^2 = V(X) = \tfrac{1}{12}(n^2 - 1).$$

Many of these discrete probability functions will be expressed subsequently in the form p_x or p_n. A simple example of the uniform discrete distribution is the fair die in which the faces $1, 2, 3, 4, 5, 6$ are equally likely to appear, each with probability $\tfrac{1}{6}$.

1.8 Some standard continuous probability distributions

The *exponential distribution* is a continuous distribution that will be used in subsequent chapters to model the random variable, say X, which is the *time* to a particular event. In a Poisson process, we shall see that this is the time between successive occurrences of the event of interest. For example, the inter-arrival time of customers in a queue, or the lifetime or the time to failure of a component, where the failure rate α is constant, in reliability theory, can be modelled by *exponential distributions*.

The density function is

$$f(x) = \alpha e^{-\alpha x}, \qquad x > 0$$

where α is a positive parameter. The mean and variance are given by

$$\mu = \mathbf{E}(X) = \int_0^\infty \alpha x e^{-\alpha x}\, \mathrm{d}x = \frac{1}{\alpha},$$

and

$$\sigma^2 = \mathbf{V}(X) = \int_0^\infty \left(x - \frac{1}{\alpha}\right)^2 \alpha e^{-\alpha x}\, \mathrm{d}x = \frac{1}{\alpha^2}.$$

It is possible to derive the distribution function $F(x)$ of this random variable: it is given by

$$F(x) = \int_0^x \alpha e^{-\alpha y}\, \mathrm{d}y = 1 - e^{-\alpha x}.$$

A closed form for $F(x)$ is not always possible for continuous random variables, and tables are sometimes necessary. Like the geometric distribution considered in the last section, the exponential distribution possesses the 'no memory' property

$$\mathbf{P}(X > a + b | X > a) = \mathbf{P}(X > b), \qquad a, b > 0 \qquad (1.5)$$

This may be proved using the definition of conditional probability in Section 1.3, and noting that $\mathbf{P}(X > x) = e^{-\alpha x}$. It can be shown that if (1.5) is true, then the random variable X has an exponential distribution.

A random variable X has a *normal distribution* with mean μ and variance $\mathbf{V}(X)$ or σ^2, if its probability density function is

$$f(x) = \frac{1}{\sigma\sqrt{2\pi}} \exp\left[-\frac{(x - \mu)^2}{2\sigma^2}\right], \qquad -\infty < x < \infty. \qquad (1.6)$$

That

$$\int_{-\infty}^\infty f(x)\, \mathrm{d}x = 1$$

follows from the standard result

$$\int_{-\infty}^{\infty} e^{-u^2} \, du = \sqrt{\pi}.$$

For the normal distribution it can be verified that the mean and variance are

$$\int_{-\infty}^{\infty} x f(x) \, dx = \mathbf{E}(X) = \mu, \qquad \int_{-\infty}^{\infty} (x - \mu)^2 f(x) \, dx = \sigma^2.$$

The normal distribution is often denoted by $N(\mu, \sigma^2)$.

The *gamma distribution* depends on the properties of the gamma function $\Gamma(n)$ defined by

$$\Gamma(n) = \int_{0}^{\infty} x^{n-1} e^{-x} \, dx.$$

Integration by parts produces the recurrence formula

$$\Gamma(n) = (n-1)\Gamma(n-1),$$

since

$$\Gamma(n) = -\int_{0}^{\infty} x^{n-1} \frac{de^{-x}}{dx} \, dx$$

$$= [-x^{n-1} e^{-x}]_{0}^{\infty} + \int_{0}^{\infty} \frac{dx^{n-1}}{dx} e^{-x} \, dx$$

$$= (n-1) \int_{0}^{\infty} x^{n-2} e^{-x} \, dx = (n-1)\Gamma(n-1)$$

When n is an integer it follows that

$$\Gamma(n) = (n-1)!.$$

The gamma distribution has two parameters n, α, and has the density function

$$f(x) = \frac{\alpha^n}{\Gamma(n)} x^{n-1} e^{-\alpha x}, \qquad x > 0.$$

Note that by setting $n = 1$, we obtain the exponential distribution. The mean is given by

$$\mathbf{E}(X) = \frac{\alpha^n}{\Gamma(n)} \int_{0}^{\infty} x^n e^{-\alpha x} \, dx = \frac{\alpha^n}{\Gamma(n)\alpha^{n+1}} \int_{0}^{\infty} y^n e^{-y} \, dy,$$

after the change of variable $y = \alpha x$. Hence

$$\mathbf{E}(X) = \frac{\Gamma(n+1)}{\alpha\Gamma(n)} = \frac{n}{\alpha}.$$

It can be shown that the variance of the gamma distribution is n/α^2. Note that the mean and variance are respectively n times the mean and variance of the exponential distribution with parameter α.

It will be proved later that if $X_1, X_2, \ldots, X_n$ are independent identically distributed exponential random variables, then

$$Y = \sum_{i=1}^{n} X_i$$

has a gamma distribution with parameters α, n.

Another distribution arising in reliability (Chapter 8) is the *Weibull distribution* which depends on two positive parameters α, β, and has density

$$f(x) = \alpha\beta x^{\beta-1} e^{-\alpha x^\beta}, \qquad x \geq 0.$$

This enables more complex lifetime data to be modelled, especially where a constant failure rate is not a reasonable assumption. Note that setting $\beta = 1$ gives the exponential distribution. By a simple change of variable to $y = x^\beta$, it can be seen that Y has an exponential distribution with parameter α. After some algebra, it may be shown that the mean and variance of the Weibull distribution are

$$\mu = \frac{1}{\alpha^{\frac{1}{\beta}}} \Gamma\left(\frac{1}{\beta}+1\right),$$

$$\sigma^2 = \frac{1}{\alpha^{\frac{2}{\beta}}} \left[\Gamma\left(\frac{2}{\beta}+1\right) - \left\{\Gamma\left(\frac{1}{\beta}+1\right)\right\}^2\right].$$

1.9 Generating functions

In this section we are going to consider two generating functions. The main purpose is to use their uniqueness properties to identify the probability distributions of functions of random variables. The first is the *moment generating function* (mgf). This function depends on a dummy variable t and uses the series expansion of e^{tx} to generate the *moments* $\mathbf{E}(X^r)$, $(r \geq 1)$ of the probability distribution of X. The expectation $\mathbf{E}(X^r)$ is called the rth moment of the random variable X about zero. Recall that

$$e^{tX} = 1 + tX + \frac{(tX)^2}{2!} + \cdots + \frac{(tX)^r}{r!} + \cdots,$$

a series which converges for all tX. The mgf is obtained by taking expectations of both sides of this equation,

$$M_X(t) = \mathbf{E}(e^{tX}) = \mathbf{E}\left[1 + tX + \frac{(tX)^2}{2!} + \cdots + \frac{(tX)^r}{r!} + \cdots\right]. \qquad (1.7)$$

We mentioned in Section 1.6 that the expectation of a finite sum of functions was equal to the sum of the expectations. We now wish to apply this result to an infinite sum, making the assumption that this result holds (it does so under fairly general conditions). Thus, we assume that

$$M_X(t) = 1 + t\mathbf{E}(X) + \frac{t^2}{2!}\mathbf{E}(X^2) + \cdots + \frac{t^r}{r!}\mathbf{E}(X^r) + \cdots = \sum_{r=0}^{\infty} \frac{t^r}{r!}\mathbf{E}(X^r).$$

The coefficient of $t^r/r!$ is therefore the rth moment of X. Taking successive derivatives of the mgf with respect to t, and then setting $t = 0$, we can obtain these moments. For example,

$$M_X'(0) = \mathbf{E}(X) = \mu,$$

$$M_X''(0) = \mathbf{E}(X^2),$$

and

$$\sigma^2 = M''(0) - [M'(0)]^2.$$

Let X have a gamma distribution with parameters n, α. Then

$$M_X(t) = \mathbf{E}(e^{tX}) = \int_0^\infty \frac{\alpha^n}{\Gamma(n)} e^{tx} x^{n-1} e^{-\alpha x} \, dx,$$

$$= \int_0^\infty \frac{\alpha^n}{\Gamma(n)} x^{n-1} e^{-x(\alpha-t)} \, dx,$$

The change of variable $x = u/(\alpha - t)$ leads to the result

$$M_X(t) = \left(\frac{\alpha}{\alpha - t}\right)^n \qquad \text{provided that } t < \alpha.$$

Now consider the result quoted but not proved in the previous section on the distribution of the sum of independent and identically distributed (iid) exponential random variables $X_1, X_2, \ldots, X_n$.

We may now use the two results above to prove this. We first need to note that, since the random variables are independent, it follows that

$$\mathbf{E}\Big(g_1(X_1)g_2(X_2)\cdots g_n(X_n)\Big) = \mathbf{E}\Big(g_1(X_1)\Big)\mathbf{E}\Big(g_2(X_2)\Big)\cdots \mathbf{E}\Big(g_n(X_n)\Big).$$

Let $Y = \sum_{i=1}^n X_i$. Then

$$M_Y(t) = \mathbf{E}(e^{tY}) = \mathbf{E}(e^{t\sum_{i=1}^n X_i})$$

$$= \mathbf{E}(\prod_{i=1}^n e^{tX_i}) = \prod_{i=1}^n \mathbf{E}(e^{tX_i})$$

$$= \prod_{i=1}^n M_{X_i}(t),$$

but since the X_is are identically distributed, they have the same mgf $M_X(t)$. Hence

$$M_Y(t) = \left[M_X(t)\right]^n.$$

Of course, this result holds for any iid random variables. If X is exponential, then

$$M_X(t) = \frac{\alpha}{\alpha - t},$$

and

$$M_Y(t) = \left(\frac{\alpha}{\alpha - t}\right)^n,$$

which is the mgf of a gamma-distributed random variable. Hence the distribution of the sum of iid exponentially distributed random variables is gamma by the uniqueness property.

Example 1.7. Find the moment generating function of the normal distribution $N(\mu, \sigma^2)$. A machine cuts rods to a nominal length of a metres. It is found that the lengths are normally distributed with mean $\mu = a$ and standard deviation $\sigma = b$ metres. (Note that the standard deviation is in the same units as the mean, unlike the variance.) The ends of pairs of rods are welded together to form a right angle. What is the expected area of the triangle formed by the rods?

If X is a random variable for the length of the rod, then, for (1.6) and (1.7),

$$M_X(t) = \mathrm{E}(e^{tX}) = \frac{1}{\sigma\sqrt{2\pi}} \int_{-\infty}^{\infty} e^{tx} \exp\left[-\frac{(x-\mu)^2}{2\sigma^2}\right] dx,$$

$$= \frac{1}{\sigma\sqrt{2\pi}} \int_{-\infty}^{\infty} \exp\left[\frac{2\sigma^2 xt - (x-\mu)^2}{2\sigma^2}\right] dx,$$

Apply the substitution $x - \mu = \sigma(v - \sigma t)$. Then

$$M_X(t) = \exp(t\mu + \frac{1}{2}\sigma^2 t^2) \int_{-\infty}^{\infty} \frac{1}{\sqrt{2\pi}} e^{-\frac{1}{2}v^2} dv$$

$$= \exp(t\mu + \frac{1}{2}\sigma^2 t^2) \times 1 = \exp(t\mu + \frac{1}{2}\sigma^2 t^2). \qquad (1.8)$$

Now

$$M_X(t) = \mathrm{E}(e^{tX}) = 1 + t\mathrm{E}(X) + \frac{1}{2}t^2\mathrm{E}(X^2) + \dots,$$

and we require $\mathrm{E}(\frac{1}{2}X^2) = \frac{1}{2}\mathrm{E}(X^2)$, since, the area of a triangle formed by two rods of lengths x_1 and x_2 is $\frac{1}{2}x_1 x_2$, since the random variables X_1 and X_2 are iid. In the expansion above for $M_X(t)$, $\frac{1}{2}\mathrm{E}(X^2)$ is the coefficient of t^2. From (1.8)

$$M_X(t) = \exp(t\mu + \frac{1}{2}\sigma^2 t^2)$$

$$= 1 + (t\mu + \frac{1}{2}\sigma^2 t^2) + \frac{1}{2!}(t\mu + \frac{1}{2}\sigma^2 t^2)^2 + \dots$$

$$= 1 + \mu t + \frac{1}{2}(\sigma^2 + \mu^2)t^2 + \dots$$

Hence $E(X^2) = \mu^2 + \sigma^2 = a^2 + b^2$ which means that the expected value of the area is $\frac{1}{2}(a^2 + b^2)$ metres^2. □

Moment generating functions may be defined for both discrete and continuous random variables but the *probability generating function* (pgf) is only defined for integer-valued random variables. To be specific, we consider the probability distribution $p_n = \mathbf{P}(N = n)$, where N can only take the values $0, 1, 2, \ldots$. (We use the notation p_n rather than $p(n)$ in this context since it conforms to the usual notation for coefficients in power series.) Again it is expressed in terms of a power series in a dummy variable s, and is defined as the expected value of s^N:

$$G_N(s) = \mathbf{E}(s^N) = \sum_{n=0}^{\infty} p_n s^n,$$

provided the right-hand side exists. The question of uniqueness arises with generating functions, since we deduce distributions from them. Later we shall represent the probability generating function by $G(s)$ without the random variable subscript. It can be shown (see Grimmett and Welsh, 1986) that two random variables X and Y have have the same probability generating function if and only if they have the same probability distributions.

In many cases the pgf will be a finite series: this will occur if the number of outcomes is finite. In others, such as in birth and death processes (Chapter 6), in which there is, theoretically, no upper bound to the population size, the series for the pgf will be infinite. However, in all cases the coefficient of s^n will give the probability p_n. If $G_N(s) = \sum_{n=0}^{\infty} p_n s^n$, and assuming that the power series converges for $0 \le s \le 1$ (remember that $p_n \ge 0$ for all n, then term-by-term differentiation gives

$$\frac{\mathrm{d}G}{\mathrm{d}s} = G'(s) = \sum_{n=1}^{\infty} n p_n s^{n-1}, \quad \frac{\mathrm{d}^2 G}{\mathrm{d}s^2} = G''(s) = \sum_{n=2}^{\infty} n(n-1) p_n s^{n-2}, \text{ for } 0 \le s \le 1.$$

The pgf has the following properties:

(a) $G_N(1) = \displaystyle\sum_{n=0}^{\infty} p_n = 1.$

(b) *mean*: $G'_N(1) = \displaystyle\sum_{n=0}^{\infty} n p_n = \mathbf{E}(N) = \mu.$

(c) *variance*: $G''_N(1) = \displaystyle\sum_{n=0}^{\infty} n(n-1) p_n = \mathbf{E}(N^2) - \mathbf{E}(N)$, so that the variance is given by

$$\mathbf{V}(N) = \sigma^2 = \mathbf{E}(N^2) - \mu^2 = G''(1) + G'(1) - [G'(1)]^2.$$

(d) $G_N^{(m)}(1) = \mathrm{d}^m G(1)/\mathrm{d}s^m = \mathbf{E}[N(N-1)\ldots(N-m+1)]$, which is called the *factorial moment* of N.

(e) If $N_1, N_2, \ldots, N_r$ are independent and identically distributed (iid) discrete random variables, and $Y = \sum_{i=1}^{r} N_i$, then

$$G_Y(s) = \mathbf{E}[s^Y] = \mathbf{E}\left[s^{\sum_{i=1}^{r} N_i}\right] = \mathbf{E}\left[\prod_{i=1}^{r} (s^{N_i})\right]$$

$$= \prod_{i=1}^{r} G_{N_i}(s),$$

and since the N_is are identically distributed

$$G_Y(s) = [G_X(s)]^r,$$

which is similar to the result for moment generating functions.

Example 1.8. The random variable N has a binomial distribution with parameters m, p. Its probability function is given by

$$p(n) = p_n = \mathbf{P}(N = n) = \binom{m}{n} p^n q^{m-n}, \qquad n = 0, 1, 2, \ldots, m$$

(see equation (1.4)). Find its pgf.

The pgf of N is

$$G_N(s) = G(s) = \sum_{n=0}^{m} s^n \binom{m}{n} p^n q^{m-n}$$

$$= \sum_{n=0}^{m} \binom{m}{n} (ps)^n q^{m-n}$$

$$= (q + ps)^m$$

using the binomial theorem. It follows that

$$G'(s) = mp(q + ps)^{m-1},$$

$$G''(s) = m(m - 1)p^2(q + ps)^{m-2}.$$

Using the results above, the mean and variance are given by

$$\mu = G'(1) = mp,$$

and

$$\sigma^2 = G''(1) + G'(1) - [G'(1)]^2,$$
$$= m(m - 1)p^2 + mp - m^2 p^2,$$
$$= mpq$$

□

The Bernoulli distribution is the binomial distribution with $n = 1$. Hence its pgf is

$$G_X(s) = q + ps.$$

Consider n independent and identically distributed Bernoulli random variables $X_1, X_2, \ldots, X_n$, and let $Y = \sum_{i=1}^{n} X_i$. Then from the results above

$$G_Y(s) = [G_X(s)]^n = (q + ps)^n,$$

which is again the pgf of a binomial random variable. (To derive this conclusion from the pgf really requires an appeal to a uniqueness result, which is not included in this text.)

It is possible to associate a generating function with any sequence $\{a_n\}$, ($n = 0, 1, 2, \ldots$) in the form

$$H(s) = \sum_{n=0}^{\infty} a_n s^n$$

provided that the series converges in some interval containing the origin $s = 0$. Unlike the pgf this series need not satisfy the conditions $H(1) = 1$ for a probability distribution nor $0 \le a_n \le 1$. An application using such a series is given in Section 3.3 on random walks.

1.10 Conditional expectation

In many applications of probability we are interested in the possible values of two or more characteristics in a problem. For this we require two or more random variables that may or may not be independent. We shall only consider the case of two discrete random variables, X and Y, which form a two-dimensional random variable denoted by (X, Y), which can take pairs of values (x_i, y_j) ($i = 1, 2, \ldots; j = 1, 2, \ldots$) (either sequence may be finite or infinite) with probability (mass) function $p(x_i, y_j)$, which is now a function of two variables. As in Section 1.4, the probabilities must satisfy

(a) $0 \le p(x_i, y_j) \le 1$ for all (i, j),

(b) $\displaystyle\sum_{i=1}^{\infty} \sum_{j=1}^{\infty} p(x_i, y_j) = 1,$

The random variables X and Y are said to be *independent*, if and only if,

$$p(x_i, y_j) = q(x_i)r(y_j) \quad \text{for all } i \text{ and } j, \tag{1.9}$$

where, of course,

$$\sum_{i=1}^{\infty} q(x_i) = 1, \quad \sum_{j=1}^{\infty} r(y_j) = 1.$$

If $Z = H(X, Y)$ is a function of X and Y, then we will state without proof that the expected value of Z is given by

$$E(Z) = \sum_{i=1}^{\infty} \sum_{j=1}^{\infty} H(x_i, y_j) p(x_i, y_j).$$

We can also consider the *conditional expectation* of X for a given value y_j of Y. This is defined as

$$E(X|Y = y_j) = \sum_{i=1}^{\infty} x_i P(X = x_i | Y = y_j) = \sum_{i=1}^{\infty} x_i p_X(x_i|y_j)$$

where $p_X(x_i|y_j)$ is the probability that $X = x_i$ occurs given that $Y = y_j$ has occurred. We can view this expectation as a function of y_i. We can also interpret it as the same function of the random variable Y: we denote this function of Y as $E(X|Y)$. Similarly

$$E(Y|X = x_i) = \sum_{j=1}^{\infty} y_j p_Y(y_j|x_i)$$

and $E(Y|X)$ can be defined. The conditional probability $p(x_i|y_j)$ is given by (see Section 1.4)

$$p_X(x_i|y_j) = \frac{p(x_i, y_j)}{p_Y(y_j)} \qquad (p_Y(y_j) \neq 0),$$

where

$$p_Y(y_j) = \sum_{i=1}^{\infty} p_X(x_i, y_j)$$

is the *marginal probability distribution* of Y.

As we have observed, the expectations $E(X|Y)$ and $E(Y|X)$ are *random variables* that are functions of Y and X respectively. However, if X and Y are independent random variables, then $p(x_i, y_j) = q(x_i) r(y_j)$ by (1.9) so that

$$p_X(x_i|y_j) = \frac{p(x_i, y_j)}{p_Y(y_j)} = \frac{q(x_i) r(y_j)}{\sum_{i=1}^{\infty} q(x_i) r(y_j)} = q(x_i),$$

and, similarly,

$$p_Y(y_i|x_j) = r(y_j).$$

Note that, for independent random variables the marginal probability distributions with respect to Y and X are $r(y_i)$ and $q(x_i)$ respectively. Hence

$$E(X|Y) = \sum_{i=1}^{\infty} x_i p_X(x_i|y_j) = \sum_{i=1}^{\infty} x_i q(x_i) = E(X).$$

By a similar argument, $E(Y|X) = E(Y)$.

On the other hand, if X and Y are dependent random variables, then $\mathbf{E}(X|Y)$ and $\mathbf{E}(Y|X)$ can themselves have expectations. Thus

$$\mathbf{E}(\mathbf{E}(X|Y)) = \sum_{j=1}^{\infty} \left(\sum_{i=1}^{\infty} x_i \frac{p(x_i, y_j)}{p_Y(y_j)} \right) p_Y(y_j),$$

since $p(y_j)$ is the probability that $Y = y_j$ occurs. Hence

$$\mathbf{E}(\mathbf{E}(X|Y)) = \sum_{j=1}^{\infty} \sum_{i=1}^{\infty} x_i \, p(x_i, y_j) = \mathbf{E}(X).$$

Similarly

$$\mathbf{E}(\mathbf{E}(Y|X)) = \mathbf{E}(Y).$$

Example 1.9. A bird lays Y eggs in a nest, where Y has a binomial distribution with parameters m and p. Each egg hatches with probability r independently of the other eggs. If X is the number of young in the nest, find the conditional probabilities $\mathbf{E}(X|Y)$ and $\mathbf{E}(Y|X)$.

Let $y_j = j$, $(0 \le j \le m)$ be possible values of Y. Since Y is binomial, using the notation above,

$$p_Y(y_j) = \binom{m}{j} p^j (1 - p)^{m-j}, \quad (0 \le j \le m).$$

Let $x_i = i$, $(0 \le i \le j)$ be possible numbers of the young in the nest. Then the conditional mass function

$$p_X(x_i|y_j) = \mathbf{P}(X = x_i|Y = y_j) = \binom{j}{i} r^i (1-r)^{j-i}, \quad (0 \le j \le m, \ 0 \le i \le j),$$

which is the number of ways in which i items can be chosen from j items. This distribution is also binomial but with parameters j and i. Hence

$$\mathbf{E}(X|Y = y_j) = \sum_{i=1}^{j} ip(x_i|y_j) = \sum_{i=1}^{j} i \binom{j}{i} r^i (1 - r)^{j-i}$$

$$= (1 - r)^j \sum_{i=1}^{j} k \binom{j}{i} \left(\frac{r}{1-r} \right)^i$$

$$= (1 - r)^j \cdot \frac{r}{1-r} \cdot j \left(1 + \frac{r}{1-r} \right)^{j-1}$$

$$= jr = y_j r.$$

Hence, $\mathbf{E}(X|Y)$ is the random variable rY, $(0 \le Y \le m)$.

To obtain the conditional expectation $\mathbf{E}(Y|X)$, start from (1.2) which can be written as

$$\mathbf{P}(Y \cap X) = \mathbf{P}(X|Y)\mathbf{P}(Y) = \mathbf{P}(X \cap Y) = \mathbf{P}(Y|X)\mathbf{P}(X).$$

Hence the conditional mass function

$$p(y_j|x_i) = \mathbf{P}(Y = y_j|X = x_i) = \frac{\mathbf{P}(X = x_i|Y = y_j)\mathbf{P}(Y = y_j)}{\mathbf{P}(X = x_i)}$$

$$= \frac{p(x_i|y_j)p(y_j)}{p_X(x_i)},$$

where

$$p_X(x_i) = \mathbf{P}(X = x_i) = \sum_{s=i}^{m} \binom{s}{i} r^i (1-r)^{s-i} \cdot \binom{m}{s} p^s (1-p)^{m-s}$$

$$= \left(\frac{r}{1-r}\right)^i (1-p)^m \sum_{s=i}^{m} \binom{s}{i}\binom{m}{s} \rho^s$$

where

$$\rho = \frac{p(1-r)}{1-p}.$$

It follows that

$$p(y_j|x_i) = \frac{\binom{j}{i}\binom{m}{j}\rho^j}{\sum_{s=i}^{m}\binom{s}{i}\binom{m}{s}\rho^s}.$$

Finally

$$\mathbf{E}(Y|X = x_i) = \frac{\sum_{j=i}^{m} j\binom{j}{i}\binom{m}{j}\rho^j}{\sum_{s=i}^{m}\binom{s}{i}\binom{m}{s}\rho^s} = u(i, r, p, m) = u(x_i, r, p, m),$$

say. Hence the conditional expectation $\mathbf{E}(Y|X)$ is the random variable $u(X, r, p, m)$. □

Problems

1.1. The Venn diagram of three events is shown in Figure 1.5. Indicate on the diagram the following events:

(a) $A \cup B$;
(b) $A \cup (B \cup C)$;
(c) $A \cap (B \cup C)$;
(d) $(A \cap C)^c$;
(e) $(A \cap B) \cup C^c$.

1.2. In a random experiment, A, B, C are three events. In set notation, write down expressions for the events:

(a) only A occurs;
(b) all three events A, B, C occur;
(c) A and B occur but C does not;

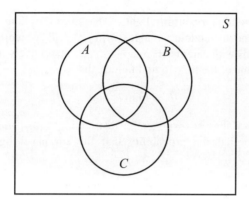

Fig. 1.5 *Venn diagram for Problem 1.1.*

(d) at least one of the events A, B, C occurs;

(e) exactly one of the events A, B, C occurs;

(f) not more than two of the events occur.

1.3. For two events A and B, $\mathbf{P}(A) = 0.4$, $\mathbf{P}(B) = 0.5$ and $\mathbf{P}(A \cap B) = 0.3$. Calculate

(a) $\mathbf{P}(A \cup B)$;

(b) $\mathbf{P}(A \cap B^c)$;

(c) $\mathbf{P}(A^c \cup B^c)$.

1.4. Two distinguishable fair dice a and b are rolled. What are the elements of the sample space? What is the probability that the sum of the face values of the two dice is 9? What is the probability that at least one 5 or at least one 3 appears?

1.5. Two distinguishable dice a and b are rolled. Construct a probability function for the sum of the face values.

1.6. A probability function $\{p_n\}$, $(n = 0, 1, 2, \ldots)$ has a probability generating function

$$G(s) = \sum_{n=0}^{\infty} p_n s^n = \tfrac{1}{4}(1 + s)(3 + s)^{\frac{1}{2}}.$$

Find p_n and its mean.

1.7. Find the probability generating function $G(s)$ of the Poisson distribution (see Section 1.7) with parameter α given by

$$p_n = \frac{e^{-\alpha}\alpha^n}{n!}, \qquad n = 0, 1, 2, \ldots.$$

Determine the mean and variance of $\{p_n\}$ from the generating function.

1.8. A panel contains n warning lights. The times to failure of the lights are the independent random variables $T_1, T_2, \ldots, T_n$ which have exponential distributions with parameters $\alpha_1, \alpha_2, \ldots, \alpha_n$ respectively. Let T be the random variable of the time to first failure, that is

$$T = \min\{T_1, T_2, \ldots, T_n\}.$$

Show that T has an exponential distribution with parameter $\sum_{j=1}^{n} \alpha_j$. Show also that the probability that the ith panel light fails first is $\alpha_i / (\sum_{j=1}^{n} \alpha_j)$.

1.9. The geometric probability function with parameter p is given by

$$p(x) = q^{x-1} p, \quad x = 1, 2, \ldots$$

where $q = 1 - p$ (see Section 1.7). Find its probability generating function. Calculate the mean and variance of the geometric distribution from its pgf.

1.10. Two distinguishable fair dice a and b are rolled. What are the probabilities that:

(a) at least one 4 appears;
(b) just one 4 appears;
(c) the sum of the face values is 6;
(d) the sum of the face values is 5 and one 3 is shown;
(e) the sum of the face values is 5 or just one 3 is shown?

1.11. Two distiguishable fair dice a and b are rolled. What is the expected sum of the face values? What is the variance of the sum of the face values?

1.12. Three distinguishable fair dice a, b and c are rolled. How many possible outcomes are there for the faces shown? When the dice are rolled, what is the probability that just two dice show the same face values and the third one is different?

1.13. In a sample space S, the events B and C are mutually exclusive, but A and B, and A and C are not. Show that

$$\mathbf{P}(A \cup (B \cup C)) = \mathbf{P}(A) + \mathbf{P}(B) + \mathbf{P}(C) - \mathbf{P}(A \cap (B \cup C)).$$

From a well-shuffled pack of 52 playing cards a single card is randomly drawn. Find the probability that it is a club or an ace or the king of hearts.

1.14. Show that

$$f(x) = \begin{cases} 0 & x < 0 \\ \frac{1}{2a} & 0 \le x \le a \\ \frac{1}{2a} e^{-(x-a)/a} & x > a \end{cases}$$

is a possible probability density function. Find the corresponding cumulative distribution function.

1.15. A biased coin is tossed. The probability of a head is p. The coin is tossed until the first head appears. Let the random variable N be the total number of tosses including the first head. Find $\mathbf{P}(N = n)$, and its pgf $G(s)$. Find the expected value of the number of tosses.

1.16. A machine cuts rods to a nominal length of a metres. It is found that the lengths are normally distributed with mean $\mu = a$ and standard deviation $\sigma = b$ metres. The ends of three of rods, selected at random, are welded together in such a way that they are mutually perpendicular. A tetrahedron is formed by joining the remaining ends. What is the expected volume of the tetrahedron formed by the rods? (See Example 1.7: the volume of a tetrahedron is $\frac{1}{3}$(base area × height).)

1.17. A random variable X has a gamma distribution with parameters n and α. Using the moment generating function of the gamma distribution, find $\mathbf{E}(X^p)$ where p is a positive integer.

1.18. A probability generating function with parameter $0 < \alpha < 1$ is given by

$$G(s) = \frac{1 - \alpha(1 - s)}{1 + \alpha(1 - s)}.$$

Find $p_n = \mathbf{P}(N = n)$ by expanding the series in powers of s. What is the mean of the probability function $\{p_n\}$?

1.19. Find the moment generating function of the random variables X which has the uniform distribution

$$f(x) = \begin{cases} 1/(b - a) & a \leq x \leq b \\ 0 & \text{for all other values of } x \end{cases}$$

Deduce $\mathbf{E}(X^n)$.

1.20. If the random variable N has the probability distribution $\{p_n\}$, $(n = 0, 1, 2 \ldots)$ and probability generating function $G(s)$, show that the random variable $Y = aX + b$ has the probability generating function

$$H(s) = s^b G(s^a),$$

where a and b are positive integers. If $p_n = 1/2^{n+1}$, $(n = 0, 1, 2, \ldots)$, determine $G(s)$ and $H(s)$, and confirm using $H(s)$ that the mean of Y is $a + b$.

1.21. Find the probability generating functions of the following distributions, in which $0 < p < 1$:

(a) Bernoulli distribution: $p_n = p^n(1-p)^n$, $(n = 0, 1)$;
(b) geometric distribution: $p_n = p(1-p)^{n-1}$, $(n = 1, 2, \ldots)$;
(c) negative binomial distribution:

$$p_n = \binom{r+n-1}{r-1} p^n (1-p)^r, \qquad (n = 0, 1, 2, \ldots)$$

where r is a positive integer. In each case find also the mean and variance of the distribution using the probability generating function.

1.22. An annual plant produces n seeds in a season which are assumed to have a Poisson distribution with parameter λ. Each seed has a probability p of germinating to create a new plant which propagates in the following year. Let N be the random variable of the number of seeds, and M the random variable of the number of new plants. Show that $E(M|N) = pN$, and that the probability function of M is

$$p_m = (p\lambda)^m e^{-p\lambda}/m!,$$

that is Poisson with parameter $p\lambda$. Find $E(N|M)$ and confirm that

$$E(E(M|N)) = E(M), \qquad E(E(N|M)) = E(N).$$

1.23. A word of five letters is transmitted by code to a receiver. The transmission signal is weak, and there is a 5% probability that any letter is in error independently of the others. What is the probability that the word is received correctly? The same word is transmitted a second time with the same errors in the signal. If the same word is received, what is the probability now that the word is correct?

1.24. A random variable N over the positive integers has the probability distribution with

$$p_n = \mathbf{P}(N = n) = -\frac{\alpha^n}{n \ln(1-\alpha)}, \qquad (0 < \alpha < 1; \; n = 1, 2, 3, \ldots).$$

What is its probability generating function? Find the mean of the random variable.

2
Some gambling problems

2.1 Gambler's ruin

Consider a game of chance between two players: A, the gambler; and B, the opponent. It is assumed that after each play, A either wins one unit from B with probability p or loses one unit to B with probability $q = 1 - p$. Similarly, B either wins from A or loses to A with probabilities q or p. The result of every play of the game is independent of the results of previous plays. The gambler A and the opponent B each start with a given number of units and the game ends when either player has lost all his or her units. What is the probability that the gambler loses all his or her money or wins all the opponents money assuming that an unlimited number of plays are possible? This is the classic *gambler's ruin* problem. In a simple example of gambler's ruin, each play could depend on the spin of a fair coin, in which case $p = q = \frac{1}{2}$.

The problem will be solved by using results from conditional probability, which then leads to a *difference equation*, and we shall have more to say about methods of solution later. There are other questions associated with this problem, such as how many plays are expected before the game finishes. In some games the player might be playing against the bank, which has a large initial stake.

2.2 Probability of ruin

The result of each play of the game is a Bernoulli random variable (Section 1.7) which can only take the values -1 and $+1$. After a series of plays, we are interested in the current capital of A. This is simply the initial capital of A plus the sum of the values of the Bernoulli random variables generated by these plays. We are also interested in how the random variable, which represents the current capital, changes or evolves with the number of plays. This is measured at discrete points when the result of each play is known.

Suppose that A has an initial capital of k units and B starts with $a - k$, where a and k are positive integers and $a > k$. If X_n is A's capital after n plays (or at time point n), then $X_0 = k$. If $X_n = 0$, then the gambler A is ruined (note that we must have $n \geq k$), whilst if $X_n = a$ ($n \geq a - k$) then B is ruined, and in both cases the game terminates. Our initial objective is the derivation of $\mathbf{P}(X_n = 0)$ for all n.

The sequence of random variables $X_0, X_1, X_2 \ldots$ represents what is known as a *random process* with a finite sample space consisting of the integers from 0 to a. These values are known as the *state* of the process at each *stage* or time point n. If C_k is the event that A is eventually ruined when starting with initial capital k, then, by using the fact that the events $X_n = 0$ $(n = k, k+1, k+2, \ldots)$, are mutually exclusive, it follows that

$$\mathbf{P}(C_k) = \sum_{n=k}^{\infty} \mathbf{P}(X_n = 0).$$

Note again that the summation starts at $n = k$ since the minimum number of steps in which the game could end must be k. Note also that the results of each trial are independent but X_n, $n = 0, 1, 2, \ldots$ are not. This is easily seen to be true by considering a particular value of X_n, say x, $(0 < x < a)$, after n plays, say. This event may only occur if previously $X_{n-1} = x-1$ or $x+1$. The state reached in any play depends on the state of the previous play only: in other words, the process is said to display the *Markov* property, of which more will be explained later.

Clearly, the calculation of $\mathbf{P}(X_n = 0)$ for all n is likely to be a long and tedious process. However, we now introduce a method for the calculation of these probabilities that avoids this: it is based on the solution of linear homogeneous difference equations.

Owing to the sequential nature of this process, after the result of a play is known, then A's capital is either increased or decreased by one unit. This capital then becomes the new capital which, in turn, becomes the initial capital for the next play. Hence, if we define $u_k = \mathbf{P}(C_k)$, then after the first play the probability of ruin is either $u_{k+1} = \mathbf{P}(C_{k+1})$ or $u_{k-1} = \mathbf{P}(C_{k-1})$. Let us consider the result of the first play, and define D to be the event that A wins and D^c the event that A loses. Using the law of total probability (Section 1.3), it follows that

$$\mathbf{P}(C_k) = \mathbf{P}(C_k|D)\mathbf{P}(D) + \mathbf{P}(C_k|D^c)\mathbf{P}(D^c). \tag{2.1}$$

Now $\mathbf{P}(C_k|D) = \mathbf{P}(C_{k+1})$ and $\mathbf{P}(C_k|D^c) = \mathbf{P}(C_{k-1})$, whilst $\mathbf{P}(D) = p$ and $\mathbf{P}(D^c) = q$, which means that equation (2.1) can be written as

$$u_k = u_{k+1}p + u_{k-1}q.$$

This equation can be re-arranged into

$$pu_{k+1} - u_k + qu_{k-1} = 0, \tag{2.2}$$

which is a second-order linear homogeneous difference equation. If the gambler starts with 0 stake then ruin is certain, whilst if the gambler starts with all the capital a, then ruin is impossible. These translate into

$$u_0 = \mathbf{P}(C_0) = 1 \quad \text{and} \quad u_a = \mathbf{P}(C_a) = 0, \tag{2.3}$$

which are the *boundary conditions* for the difference equation (2.2).

It is possible to enumerate the solution (see Tuckwell, 1988): this will also be the subject of Problem 2.17. However, here we shall describe a method that also has a more general application (see Jordan and Smith, 1997, for a more detailed explanation of the solution of difference equations). Consider a solution of the form

$$u_k = Am^k,$$

where A is a constant and m is so far undefined. Direct substitution into the left-hand side of equation (2.2) yields

$$pu_{k+1} - u_k + qu_{k-1} = Am^{k-1}(pm^2 - m + q).$$

This is zero (assuming that $A \neq 0$) if either $m = 0$, which is known as the trivial solution and is not usually of interest in this context, or if m satisfies the quadratic equation

$$pm^2 - m + q = (pm - q)(m - 1) = 0, \qquad (p + q = 1),$$

which is known as the *characteristic equation* of the difference equation (2.2). The roots of the equation can be defined as $m_1 = 1$ and $m_2 = q/p$. Since the difference equation is linear, the general solution is – provided that $p \neq q$ – any linear combination of the two solutions with $m = m_1$ and $m = m_2$, that is

$$u_k = A_1 m_1^k + A_2 m_2^k = A_1 + A_2 \left(\frac{q}{p}\right)^k,$$

where A_1 and A_2 are arbitrary constants The boundary conditions $u_0 = 1$ and $u_a = 0$ imply that

$$A_1 + A_2 = 1 \qquad \text{and} \qquad A_1 + A_2 s^a = 0,$$

or

$$A_1 = -\frac{s^a}{1 - s^a} \qquad \text{and} \qquad A_2 = \frac{1}{1 - s^a},$$

where $s = q/p$. Hence, the probability that the gambler is ruined given an initial capital of k is, if $p \neq \frac{1}{2}$,

$$u_k = \frac{s^k - s^a}{1 - s^a}. \tag{2.4}$$

The special case $p = q = \frac{1}{2}$ has to be treated separately. The characteristic equation of

$$\tfrac{1}{2}u_{k+1} - u_k + \tfrac{1}{2}u_{k-1} = 0$$

is

$$m^2 - 2m + 1 = 0,$$

which has the repeated root $m_1 = m_2 = 1$. In this case one solution is 1, but we still require a second independent solution. For a repeated root of difference equations we try $k \times$ the repeated root. Hence

$$\tfrac{1}{2}u_{k+1} - u_k + \tfrac{1}{2}u_{k-1} = \tfrac{1}{2}(k+1) - k + \tfrac{1}{2}(k-1) = 0.$$

Thus, the general solution is

$$u_k = A_1 + A_2 k.$$

With the same boundary conditions, it follows that

$$u_k = \frac{a-k}{a}. \tag{2.5}$$

If the game is now looked at from B's viewpoint, then to obtain his/her probability of ruin, v_{a-k}, say, there is no need to derive this from first principles. In the above derivation, simply interchange p and q, and replace k by $a - k$ in equations (2.4) and (2.5). Hence B is ruined with probability

$$v_{a-k} = \left(\frac{1}{s^{a-k}} - \frac{1}{s^a}\right) \Big/ \left(1 - \frac{1}{s^a}\right) = \frac{s^k - 1}{s^a - 1},$$

if $s \neq 1$, and with probability

$$v_{a-k} = \frac{k}{a},$$

if $s = 1$. In both cases it follows that

$$u_k + v_{a-k} = 1.$$

Hence, the game must terminate eventually with one of the players losing.

Example 2.1. Suppose the possibility of a draw is now included. Let the probability that the gambler wins, loses or draws against the opponent in a play be respectively p, q or r. Since these are the only possible outcomes $p + q + r = 1$. Show that the probability of ruin u_k is given by

$$u_k = \frac{s^k - s^a}{1 - s^a}, \qquad s = \frac{q}{p}, \qquad p \neq \frac{1}{2}.$$

where the gambler's initial stake is k and the total at stake is a

In this case, the law of total probability given by (2.1) is extended to include a third possible outcome. In the draw, the stakes remain unchanged after the play. Hence, the probability of ruin u_k satisfies

$$u_k = u_{k+1}p + u_k r + u_{k-1}q,$$

or

$$pu_{k+1} - (1 - r)u_k + qu_{k-1} = 0. \tag{2.6}$$

As in the main problem, the boundary conditions are $u_0 = 1$ and $u_a = 0$. Replace $1 - r$ by $p + q$ in the difference equation. Its characteristic equation becomes

$$pm^2 - (p + q)m + q = (pm - q)(m - 1) = 0,$$

which has the general solution

$$u_k = A_1 + A_2 \left(\frac{q}{p}\right)^k.$$

Notice that this is the same solution as the standard problem with no draws, q/p is still the ratio of the probability of losing to winning at each play. The only difference is that $p + q \neq 1$ in this case. Hence, as in (2.4)

$$u_k = \frac{s^k - s^a}{1 - s^a},$$

where $s = q/p$. The presence of the draw extends the length of the game compared with the game with no draw and the same k, a and s. □

Example 2.2. In a gambler's ruin game the initial stakes are k and $a - k$ for the two players: if the gambler wins then s/he wins two units but if the gambler loses then he or she loses one unit. The probability of win or lose at each play is $p = \frac{1}{3}$ or $1 - p = \frac{2}{3}$. What are the boundary conditions and what is the probability of ruin?

If u_k is the probability of the gambler's ruin, then the law of total probability has to be changed in this problem to

$$u_k = pu_{k+2} + (1 - p)u_{k-1},$$

for $1 \leq k \leq a - 2$.

The boundary conditions $u_0 = 1$ and $u_a = 0$ still hold. If $k = a - 2$, then the law of total probability has to be amended to

$$u_{a-1} = pu_a + (1 - p)u_{a-2},$$

or

$$u_{a-1} - (1 - p)u_{a-2} = 0,$$

and this becomes the third boundary condition. Three boundary conditions are required since the difference equation for u_k is now the *third-order* equation

$$pu_{k+2} - u_k + (1 - p)u_{k-1} = 0.$$

Its characteristic equation is, with $p = \frac{1}{3}$,

$$m^3 - 3m + 2 = (m - 1)^2(m + 2) = 0,$$

which has the roots $1, 1, -2$. Hence the general solution is (note the repeated root)

$$u_k = A_1 + A_2 k + A_3(-2)^k.$$

The three boundary conditions above lead to

$$A_1 + A_3 = 1,$$

$$A_1 + A_2 a + A_3(-2)^a = 0,$$

$$A_1 + A_2(a - 1) + A_3(-2)^{a-1} = \frac{2}{3}(A_1 + A_2(a - 2) + A_3(-2)^{a-2}),$$

for A_1, A_2 and A_3. The solutions of these linear equations are (after some algebra)

$$A_1 = 1 - A_3 = \frac{(1 + 3a)(-2)^a}{(1 + 3a)(-2)^a - 1}, \qquad A_2 = \frac{-3(-2)^a}{(1 + 3a)(-2)^a - 1},$$

from which it follows that

$$u_k = \frac{(1 + 3a - 3k)(-2)^a - (-2)^k}{(1 + 3a)(-2)^a - 1}.$$

$\square$

2.3 Some numerical simulations

It is very easy now with personal computers to simulate probability problems with simple programs. Figure 2.1 shows a simulation of the gambler's ruin in the case $a = 20, k = 10$ and $p = \frac{1}{2}$. The program runs until the gambler's stake is either 0 or 20 in this example. Some sample probabilities are shown in Figure 2.2 again with $a = 20$ for various initial probabilities and play probabilities p. So, for example, the probability of ruin with $p = 0.42$ and initial stake of 17 is about 0.62. The figure shows that, in a game in which both parties start with the same initial stake 10, the gambler's probability of ruin becomes very close to certainty for any p less than 0.35. An alternative view is shown in Figure 2.3, which shows the value of u_{10} versus p. The figure emphasizes that there is only a real contest between the players if p lies between 0.4 and 0.6.

2.4 Expected duration of the game

It is natural in the gambler's ruin problem to be curious about how long, or really how many plays, we would expect the game to last. This is the number of steps to termination in which either the gambler or the opponent loses the game. Let

gambler's stake

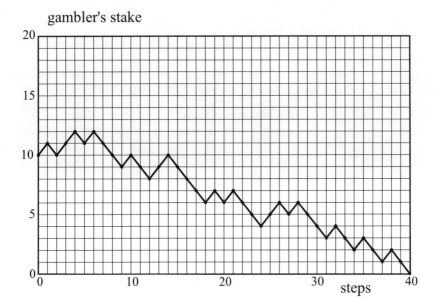

Fig. 2.1 *The figure shows the steps in a simulation of a gambler's ruin problem with* $a = 20$ *and* $k = 10$ *with probability* $p = \frac{1}{2}$ *at each play. Ruin occurs after 40 steps in this case, although by (2.5) the gambler has even chances of winning or losing the game.*

u_k probability of ruin

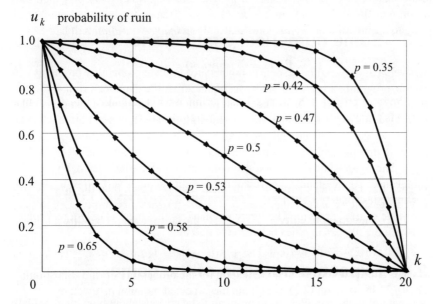

Fig. 2.2 *The ruin probability* $u_k = (s^k - s^a)/(1 - s^a)$, $(p \neq \frac{1}{2})$, $u_k = (a-k)/a$, $(p = \frac{1}{2})$ *versus* k *for* $a = 20$ *and a sample of probabilities* p.

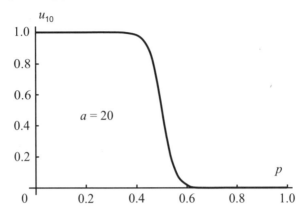

Fig. 2.3 *The probability u_{10} is shown against the play probability p for the case $a = 20$.*

us first consider a situation where the state of the game is defined in terms of two variables: k, the initial capital; and n, the remaining number of plays until the end of the game. Now n is unknown, and is a value of the random variable N, which depends, in turn, on the results of the remaining plays.

Let $p(n|k)$ be the conditional probability that the game ends in n steps given that the initial capital is k. Clearly, n will be any positive integer greater or equal to the smaller of k and $a - k$ since if the gambler won (lost) *every* play then s/he would win (lose) the game in $a - k$ (k) plays. Hence, the expected number of plays to termination or, as it is known also, the *expected duration* will be

$$\mathbf{E}(N) = \sum_{n=0}^{\infty} np(n|k) = d_k, \tag{2.7}$$

say. We have proved in Section 2.2 that termination is certain eventually so that $p(n|k)$ is a probability function and must therefore satisfy

$$\sum_{n=0}^{\infty} p(n|k) = 1,$$

for each fixed k. After the result of the next play is known then the process will move from state (k, n) to either state $(k + 1, n - 1)$ with probability p, or to state $(k - 1, n - 1)$ with probability $q = 1 - p$. By the law of total probability it follows that

$$p(n|k) = p(n - 1|k + 1)p + p(n - 1|k - 1)q, \qquad n, k \geq 1,$$

which is similar to the difference equation for the probability of ruin. Substituting for $p(n|k)$ in equation (2.7), we obtain the expected duration d_k given by

$$d_k = p \sum_{n=1}^{\infty} np(n - 1|k + 1) + q \sum_{m=1}^{\infty} mp(n - 1|k - 1).$$

The change of variable $r = n - 1$ in both summations leads to

$$d_k = p \sum_{r=0}^{\infty} (r+1)p(r|k+1) + q \sum_{r=0}^{\infty} (r+1)p(r|k-1),$$

$$= p \sum_{r=0}^{\infty} rp(r|k+1) + q \sum_{r=0}^{\infty} rp(r|k-1) + p \sum_{r=0}^{\infty} p(r|k+1) + q \sum_{r=0}^{\infty} p(r|k-1),$$

$$= p \sum_{r=0}^{\infty} rp(r|k+1) + q \sum_{r=0}^{\infty} rp(r|k-1) + p + q,$$

since, for the probability functions $p(r|k+1)$ and $p(r|k-1)$,

$$\sum_{r=0}^{\infty} p(r|k+1) = \sum_{r=0}^{\infty} p(r|k-1) = 1.$$

Hence the expected duration d_k satisfies the difference equation

$$d_k = pd_{k+1} + qd_{k-1} + 1,$$

since $p + q = 1$ and

$$d_{k+1} = \sum_{r=0}^{\infty} rp(r|k+1) \quad \text{and} \quad d_{k-1} = \sum_{r=0}^{\infty} rp(r|k-1).$$

This equation can be re-arranged into

$$pd_{k+1} - d_k + qd_{k-1} = -1, \tag{2.8}$$

which is similar to the difference equation for the probability u_k except for the term on the right-hand side. This is now a linear inhomogeneous second-order difference equation. The boundary conditions are again obtained by considering the extremes where the players lose. Thus, if $k = 0$ or $k = a$, then the game terminates so that the expected durations must be zero, that is

$$d_0 = d_a = 0.$$

The overall solution to this type of difference equation is the sum of the general solution of the corresponding *homogeneous* equation and a *particular* solution of the complete equation. The homogeneous equation has, for $s \neq 1$, the general solution

$$A_1 + A_2 s^k,$$

and, for $s = 1$, the general solution

$$A_1 + A_2 k.$$

For the particular solution, we look at the right-hand side and try a suitable function. Since any constant is a solution of the homogeneous equation, we try $d_k = Ck$, where C is a constant. Thus,

$$pd_{k+1} - d_k + qd_{k-1} + 1 = pC(k+1) - Ck + qC(k-1) + 1 = C(p-q) + 1 = 0,$$

for all k if $C = 1/(q - p)$. Hence the full general solution is

$$d_k = A_1 + A_2 s^k + \frac{k}{q - p}.$$

The boundary conditions imply

$$A_1 + A_2 = 0, \quad \text{and} \quad A_1 + A_2 s^a + \frac{a}{q - p} = 0.$$

Hence

$$A_1 = -A_2 = -\frac{a}{(q - p)(1 - s^a)},$$

with the result that the expected duration for $s \neq 1$ is

$$d_k = -\frac{a(1 - s^k)}{(q - p)(1 - s^a)} + \frac{k}{q - p} = \frac{1}{1 - 2p}\left[k - \frac{a(1 - s^k)}{1 - s^a}\right]. \qquad (2.9)$$

If $s = 1$, the difference equation (2.8) becomes

$$d_{k+1} - 2d_k + d_{k-1} = -2.$$

The general solution of the homogeneous equation is

$$A_1 + A_2 k,$$

which means that we must try Ck^2 now for the particular solution. Thus

$$d_{k+1} - 2d_k + d_{k-1} + 2 = C(k+1)^2 - 2Ck^2 + C(k-1)^2 + 2 = 2C + 2 = 0,$$

if $C = -1$. Hence

$$d_k = A_1 + A_2 k - k^2 = k(a - k), \qquad (2.10)$$

using the boundary conditions.

A sample of expected durations is shown in Figure 2.4 for a total stake of $a = 20$ and different probabilities p. For $p = \frac{1}{2}$ the expected duration has a maximum number of 100 when $k = 10$. On average, a game in which both players start with 10 units each will last for 100 plays.

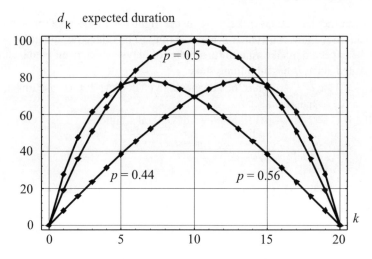

d_k expected duration

Fig. 2.4 *Expected duration d_k against k for $a = 20$ and a selection of play probabilities p.*

2.5 Some variations of gambler's ruin

2.5.1 The infinitely rich opponent

Consider the gambler's ruin problem in which the opponent is assumed to be infinitely rich. This models a gambler playing against the bank or a casino whose resources are very large. As before, the gambler's initial stake is a finite integer k, but effectively the bank's resources are infinite, that is, $a = \infty$. Hence, we can examine the limits of the gambler's probability of ruin, u_k and the expected duration, d_k as $a \to \infty$. The result depends on magnitude of s as we might expect.

(a) $s < 1$ or $p > \frac{1}{2}$. The gambler A has the advantage in each play. Now

$$\lim_{a \to \infty} s^a = 0.$$

Hence, from equation (2.4)

$$\lim_{a \to \infty} \frac{s^k - s^a}{1 - s^a} = s^k < 1.$$

The expected number of plays until this happens is given by equation (2.9), which states that

$$d_k = -\frac{a(1 - s^k)}{(q - p)(1 - s^a)} + \frac{k}{q - p}.$$

Here, $s^a \to 0$, since $s < 1$: hence a in the first term dominates and $\lim_{a\to\infty} d_k = \infty$. Hence A is not certain to be ruined but the game would be expected to take an infinite number of plays, which itself is to be anticipated since B has an infinite stake.

(b) $s > 1$ or $p < \frac{1}{2}$. B has the advantage in each play. In this case $\lim_{a\to\infty} s^a = \infty$, so that, from (2.4),

$$\lim_{a\to\infty} u_k = \lim_{a\to\infty} \frac{(s^k/s^a) - 1}{(1/s^a) - 1} = 1,$$

and

$$\lim_{a\to\infty} d_k = \lim_{a\to\infty} \frac{-a}{s^a} \frac{(1 - s^k)}{(\frac{1}{s^a} - 1)(q - p)} + \frac{k}{q - p} = \frac{k}{q - p},$$

since

$$\lim_{a\to\infty} \frac{a}{s^a} = 0.$$

Ruin is certain in an expected duration of $k/(q - p)$.

(c) $s = 1$ or $p = \frac{1}{2}$. For the case of equal odds,

$$\lim_{a\to\infty} u_k = \lim_{a\to\infty} \left(1 - \frac{k}{a}\right) = 1, \quad \text{and} \quad \lim_{a\to\infty} d_k = \lim_{a\to\infty} k(a - k) = \infty.$$

Ruin is certain but it may take a great deal of time. This is the game for someone who enjoys playing, and does not want to go to the expense of buying a biased coin.

2.5.2 The generous gambler

Suppose that both players start with finite capital, and suppose that whenever A loses his/her last unit, one unit is returned to A so that A is never ruined. The opponent B is the generous gambler. As a consequence we might expect that B must be ruined since A cannot lose. Hence there can be no possibility, other than $u_k = 0$. This can be checked by solving (2.2) subject to the boundary conditions

$$u_0 = u_1, \quad \text{and } u_a = 0.$$

For the expected duration, the boundary condition at $k = 0$ must be modified. Since one unit is returned, the expected duration at $k = 0$ must be the same as that at $k = 1$. The boundary conditions become

$$d_0 = d_1, \quad \text{and} \quad d_a = 0.$$

Consider the case of equal odds at each play. Then, for $p = \frac{1}{2}$,

$$d_k = A_1 + A_2 k - k^2.$$

Hence $A_1 = a^2 - a$ and $A_2 = 1$, so that

$$d_k = (a - k)(a + k - 1).$$

There are thus $(a - k)(a - 1)$ more plays than in the standard game.

2.5.3   Changing the stakes

Suppose that in the original game the stakes per play are halved for both players so that in the new game A has effectively $2k$ units and B has $2(a - k)$ units. How is the probability of ruin changed?

Let v_k be the probability of ruin. Then, by analogy with the formula for u_k given by equation (2.4),

$$v_k = \frac{s^{2k} - s^{2a}}{1 - s^{2a}} = \frac{(s^k - s^a)(s^k + s^a)}{(1 - s^a)(1 + s^a)} = u_k \frac{s^k - s^a}{1 + s^a}, \qquad s \neq 1.$$

If $s < 1$, then

$$\frac{s^k + s^a}{1 + s^a} < 1$$

so that $v_k < u_k$. On the other hand if $s > 1$ then

$$\frac{s^k + s^a}{1 + s^a} > 1$$

so that $v_k > u_k$. If $s < 1$ that is $p > \frac{1}{2}$, then it could be wise for A to agree to this change of play. As might be expected for $s = 1$, that is equal odds with $p = \frac{1}{2}$, the probability of ruin is unaffected.

Problems

2.1. In the standard gambler's ruin problem with total stake a and gambler's stake k, and the gambler's probability of winning at each play is p, calculate the probability of ruin in the following cases;

(a) $a = 100$, $k = 5$, $p = 0.6$;
(b) $a = 80$, $k = 70$, $p = 0.45$;
(c) $a = 50$, $k = 40$, $p = 0.5$.

Also find the expected duration in each case.

2.2. In a casino game based on the standard gambler's ruin, the gambler and the dealer each start with 20 tokens and one token is bet on at each play. The game continues until one player has no further tokens. It is decreed that the probability that any gambler is ruined is 0.52 to protect the casino's profit. What should the probability that the gambler wins at each play be?

2.3. Find general solutions of the following difference equations:

(a) $u_{k+1} - 4u_k + 3u_{k-1} = 0$;
(b) $7u_{k+2} - 8u_{k+1} + u_k = 0$;
(c) $u_{k+1} - 3u_k + u_{k-1} + u_{k-2} = 0$.
(d) $pu_{k+2} - u_k + (1 - p)u_{k-1} = 0$, $(0 < p < 1)$.

2.4. Solve the following difference equations subject to the given boundary conditions:

(a) $u_{k+1} - 6u_k + 5u_{k-1} = 0$, $\quad u_0 = 1, u_4 = 0$;

(b) $u_{k+1} - 2u_k + u_{k-1} = 0$, $\quad u_0 = 1, u_{20} = 0$;

(c) $d_{k+1} - 2d_k + d_{k-1} = -2$, $\quad d_0 = 0, d_{10} = 0$;

(d) $u_{k+2} - 3u_k + 2u_{k-1} = 0$, $\quad u_0 = 1, u_{10} = 0, 3u_9 = 2u_8$.

2.5. Show that a difference equation of the form

$$au_{k+2} + bu_{k+1} - u_k + cu_{k-1} = 0,$$

where $a, b, c \geq 0$ are probabilities with $a + b + c = 1$, can never have a characteristic equation with complex roots.

2.6. In the standard gambler's ruin problem, with equal probabilities $p = q = \frac{1}{2}$, find the expected duration of the game given the usual initial stakes of k units for the gambler and $a - k$ units for the opponent.

2.7. In a gambler's ruin problem, the possibility of a draw is included. Let the probability that the gambler wins, loses or draws against an opponent be, respectively, $p, p, 1 - 2p$, $(0 < p \leq \frac{1}{2})$. Find the probability that the gambler loses the game given the usual initial stakes of k units for the gambler and $a - k$ units for the opponent. Show that d_k, the expected duration of the game, satisfies

$$pd_{k+1} - 2pd_k + pd_{k-1} = -1.$$

Solve the difference equation and find the expected duration of the game.

2.8. In the changing stakes game in which a game is replayed with each player having twice as many units, $2k$ and $2(a - k)$ respectively, suppose that the probability of a win for the gambler at each play is $\frac{1}{2}$. Whilst the probability of ruin is unaffected, by how much is the expected duration of the game extended compared with the original game?

2.9. A roulette wheel has 37 radial slots of which 18 are red, 18 are black and 1 is green. The gambler bets one unit on either red or black. If the ball falls into a slot of the same colour, then the gambler wins one unit, and if the ball falls into the other colour (red or black), then the casino wins. If the ball lands in the green slot, then the bet remains for the next spin of the wheel, or more if necessary, until the ball lands on a red or black. The original bet is either *returned* or lost depending on whether the outcome matches the original bet or not (this is the Monte Carlo system). Show that the probability u_k of ruin for a gambler who starts with k chips with the casino holding $a - k$ chips satisfies the difference equation

$$36u_{k+1} - 73u_k + 37u_{k-1} = 0.$$

Solve the difference equation for u_k. If the house starts with Ff1,000,000 at the roulette wheel and the gambler starts with Ff10,000, what is the probability that the gambler breaks the bank if Ff5,000 are bet at each play.

In the US system the rules are less generous to the players. If the ball lands on green then the player simply loses. What is the probability now that the player wins given the same initial stakes? (see Luenberger, 1979)

2.10. In a single trial the possible scores 1 and 2 can occur each with probability $\frac{1}{2}$. If p_n is the probability of scoring *exactly* n points at some stage, show that

$$p_n = \tfrac{1}{2} p_{n-1} + \tfrac{1}{2} p_{n-2}.$$

Calculate p_1 and p_2, and find a formula for p_n. How does p_n behave as n becomes large? How do you interpret the result?

2.11. In a single trial the possible scores 1 and 2 can occur with probabilities α and $1 - \alpha$, $0 < \alpha < 1$. Find the probability of scoring exactly n points at some stage in an indefinite succession of trials. Show that

$$p_n \to \frac{1}{2 - \alpha},$$

as $n \to \infty$.

2.12. The probability of success in a single trial is $\frac{1}{3}$. If u_n is the probability that there are no two consecutive successes in n trials, show that u_n satisfies

$$u_{n+1} = \tfrac{2}{3} u_n + \tfrac{2}{9} u_{n-1}.$$

What are the values of u_1 and u_2? Hence show that

$$u_n = \frac{1}{6} \left[(3 + 2\sqrt{3}) \left(\frac{1 + \sqrt{3}}{3} \right)^n + (3 - 2\sqrt{3}) \left(\frac{1 - \sqrt{3}}{3} \right)^n \right].$$

2.13. A gambler with initial capital k units plays against an opponent with initial capital $a - k$ units. At each play of the game the gambler either wins one unit or loses one unit with probability $\frac{1}{2}$. Whenever the opponent loses the game, the gambler returns one unit so that the game may continue. Show that the expected duration of the game is $k(2a - 1 - k)$ plays.

2.14. In the usual gambler's ruin problem, the probability that the gambler is eventually ruined is

$$u_k = \frac{s^k - s^a}{1 - s^a}, \qquad s = \frac{1 - p}{p}, \qquad (p \neq \tfrac{1}{2}).$$

In a new game the stakes are halved, whilst the players start with the same initial sums. How does this affect the probability of losing by the gambler?

Should the gambler agree to this change of rule if $p < \frac{1}{2}$? By how many plays is the expected duration of the game extended compared with the original games?

2.15. In a gambler's ruin game, suppose that the gambler can win £2 with probability $\frac{1}{3}$ or lose £1 with probability $\frac{2}{3}$. Show that

$$u_k = \frac{(3k - 1 - 3a)(-2)^a + (-2)^k}{1 - (3a + 1)(-2)^a}.$$

Compute u_k if $a = 9$ for $k = 1, 2, \ldots, 8$.

2.16. Find the general solution of the difference equation

$$u_{k+2} - 3u_k + 2u_{k-1} = 0.$$

A reservoir with total capacity of a volume units of water has, during each day, either a net inflow of two units with probability $\frac{1}{3}$ or a net outflow of one unit with probability $\frac{2}{3}$. If the reservoir is full or nearly full, any excess inflow is lost in an overflow. Derive a difference equation for this model for u_k, the probability that the reservoir will eventually become empty given that it initially contains k units. Explain why the upper boundary conditions can be written

$$u_a = u_{a-1} \quad \text{and} \quad u_a = u_{a-2}.$$

Show that the reservoir is certain to be empty at some time in the future.

2.17. Consider the standard gambler's ruin problem in which the total stake is a and gambler's stake is k, and the gambler's probability of winning at each play is p and losing is $q = 1 - p$. Find u_k, the probability of the gambler losing the game, by the following alternative method. List the difference equation (2.2) as

$$u_2 - u_1 = s(u_1 - u_0) = s(u_1 - 1)$$
$$u_3 - u_2 = s(u_2 - u_1) = s^2(u_1 - 1)$$
$$\vdots$$
$$u_k - u_{k-1} = s(u_{k-1} - u_{k-2}) = s^{k-1}(u_1 - 1),$$

where $s = q/p \neq \frac{1}{2}$ and $k = 2, 3, \ldots a$. The boundary condition $u_0 = 1$ has been used in the first equation. By adding the equations show that

$$u_k = u_1 + (u_1 - 1)\frac{s - s^k}{1 - s}.$$

Determine u_1 from the other boundary condition $u_a = 0$, and hence find u_k. Adapt the same method for the special case $p = q = \frac{1}{2}$

2.18. A car park has 100 parking spaces. Cars arrive and leave randomly. Arrivals or departures of cars are equally likely, and it is assumed that simultaneous events have negligible probability. The 'state' of the car park changes whenever a car arrives or departs. Given that at some instant there are k cars in the car park, let u_k be the probability that the car park first becomes full before it becomes empty. What are the boundary conditions for u_0 and u_{100}? How many car movements can be expected before this occurs?

2.19. In a standard gambler's ruin problem with the usual parameters, the probability that the gambler loses is given by

$$u_k = \frac{s^k - s^a}{1 - s^a}, \qquad s = \frac{1-p}{p}.$$

If p is close to $\frac{1}{2}$, given, say, by $p = \frac{1}{2} + \varepsilon$ where $|\varepsilon|$ is small, show, by using binomial expansions, that

$$u_k = \frac{a-k}{a}\left[1 - 2k\varepsilon - \frac{4}{3}(a-2k)\varepsilon^2 + O(\varepsilon^3)\right],$$

as $\varepsilon \to 0$. [The O teminology is defined as follows. We say that a function $g(\varepsilon) = O(\varepsilon^b)$ as $\varepsilon \to 0$ if $g(\varepsilon)/\varepsilon^b$ is bounded in a neighbourhood which contains $\varepsilon = 0$.]

2.20. A gambler plays a game against a casino according to the following rules. The gambler and casino each start with 10 chips. From a deck of 53 playing cards which includes a joker, cards are randomly and successively drawn with replacement. If the card is red or the joker the casino wins 1 chip from the gambler, and if the card is black the gambler wins 1 chip from the casino. The game continues until either player has no chips. What is the probability that the gambler wins? What will be the expected duration of the game?

2.21. In a standard gambler's ruin problem, $a = 4$ and $k = 2$ and the probability that the gambler wins at any play is p and loses is $q = 1 - p$. If the random variable X_n is the gambler's capital after n plays (remember that the game terminates if, for any n, $X_n = a$), show that

$$\mathbf{P}(X_{2n} = 0) = q^2(2pq)^{n-1}, \qquad \mathbf{P}(X_{2n-1} = 0) = 0, \qquad (n = 1, 2, \ldots).$$

Hence show that the probability that the gambler loses is

$$\sum_{n=2}^{\infty} \mathbf{P}(X_n = 0) = \frac{q^2}{1 - 2pq}.$$

Confirm that this answer agrees with the formula for u_2 given by equation (2.4).

2.22. In the standard gambler's ruin problem with total stake a and gambler's stake k, the probability that the gambler loses is

$$u_k = \frac{s^k - s^a}{1 - s^a},$$

where $s = (1 - p)/p$. Suppose that $u_k = \frac{1}{2}$, that is fair odds. Express k as a function of a. Show that, if $s > 1$, then

$$k \approx a - \frac{\ln 2}{\ln s},$$

for large a. What is the approximate formula for $s < 1$ and large a?

2.23. In a gambler's ruin game the probability that the gambler wins at each play is α_k and loses is $1 - \alpha_k$, $(0 < \alpha_k < 1, \ 0 \leq k \leq a - 1)$, that is, the probability varies with the current stake. The probability u_k that the gambler eventually loses satisfies

$$u_k = \alpha_k u_{k+1} + (1 - \alpha_k)u_{k-1}, \quad u_o = 1, \quad u_a = 0.$$

Suppose that u_k is a specified function such that $0 < u_k < 1$, $(1 \leq k \leq a - 1)$, $u_0 = 1$ and $u_a = 0$. What should α_k be in terms of u_{k-1}, u_k and u_{k+1}?

Find α_k in the following cases:

(a) $u_k = (a - k)/a$;
(b) $u_k = (a^2 - k^2)/a^2$;
(c) $u_k = \frac{1}{2}[1 + \cos(k\pi/a)]$.

2.24. In a gambler's ruin game, the probability that the gambler wins at each play is α_k and loses is $1 - \alpha_k$, $(0 < \alpha_k < 1, \ 1 \leq k \leq a - 1)$, that is, the probability varies with the current stake. The probability u_k that the gambler eventually loses satisfies

$$u_k = \alpha_k u_{k+1} + (1 - \alpha_k)u_{k-1}, \quad u_0 = 1, \quad u_a = 0.$$

Reformulate the difference equation as

$$u_{k+1} - u_k = \beta_k(u_k - u_{k-1}),$$

where $\beta_k = (1 - \alpha_k)/\alpha_k$. Hence show that

$$u_k = u_1 + \gamma_{k-1}(u_1 - 1), \quad (k = 2, 3, \ldots, a)$$

where

$$\gamma_k = \beta_1 + \beta_1\beta_2 + \cdots + \beta_1\beta_2 \ldots \beta_k.$$

Using the boundary condition at $k = a$, confirm that

$$u_k = \frac{\gamma_{a-1} - \gamma_{k-1}}{1 + \gamma_{a-1}}.$$

Check that this formula gives the usual answer if $\alpha_k = p \neq \frac{1}{2}$, a constant.

2.25. Suppose that a fair n-sided die is rolled n independent times. A match is said to occur if side i is observed on the ith trial, where $i = 1, 2, \ldots, n$.

(a) Show that the probability of at least one match is

$$1 - \left(1 - \frac{1}{n}\right)^n.$$

(b) What is the limit of this probability as $n \to \infty$?
(c) What is the probability that just one match occurs in n trials?
(d) What is the probability that two or more matches occur in n trials?

3
Random walks

3.1 Introduction

Another way of modelling the gambler's ruin problem of the last chapter is as a *random walk* along a straight line in the following manner. Suppose that $a + 1$ positions are marked out on a straight line and numbered $0, 1, 2, \ldots, a$. A person starts at k where $0 < k < a$. The walk proceeds in such a way that, at each step, there is a probability p that the walker goes 'forward' one pace to $k + 1$, and a probability $q = 1 - p$ that the walker goes 'back' one pace to $k - 1$. The walk continues until either 0 or a is reached, and then ends. Generally, in a random walk, the position of a walker after having moved n times is known as the *state* of the walk after n *steps* or after covering n *stages*. Thus, the walk described above starts at stage k at step 0 and moves to either stage $k - 1$ or stage $k + 1$ after 1 step, and so on.

If the walk is bounded, then the ends of the walk are known as *barriers*, and they may have various properties. In this case, the barriers are said to be *absorbing*, which implies that the walk must end once a barrier is reached since there is no escape. A useful diagrammatic way of representing random walks is by a *transition* or *process diagram* as shown in Figure 3.1. In a transition diagram, the possible states of the walker can be represented by points on a line. If a transition between two points can occur in one step then those points are joined by a curve or *edge* as shown with an arrow indicating the direction of the walk, and a *weighting* denoting the probability of the step occurring. In discrete mathematics or graph theory, the transition diagram is known as a *directed graph*. A walk in the transition diagram is a succession of edges covered without a break. In

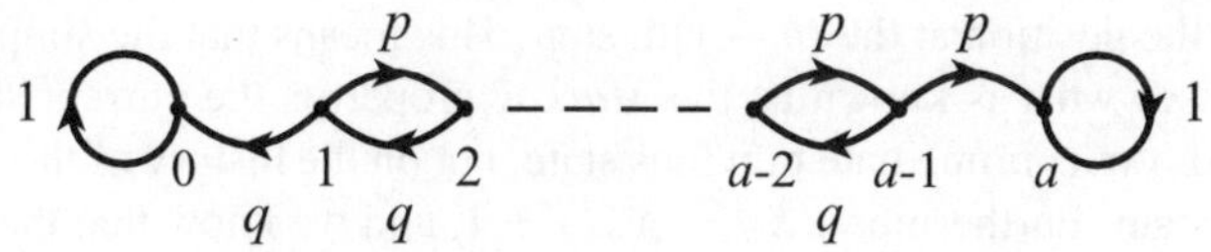

Fig. 3.1 *Transition diagram for gambler's ruin.*

Figure 3.1 the closed loops with weightings of 1 at the ends of the walk indicate the absorbing barriers with no escape.

In Section 2.5.1 we met a variant of the gambler's ruin problem. The case of the infinitely rich opponent is a random walk with an absorbing barrier at 0 but is unrestricted above 0, so that it becomes a semi-infinite random walk.

The generous opponent encountered in section 2.5.2 has an absorbing barrier at *a* but a *reflecting barrier* at 0. After reaching 0 with probability $1 - p$ the walker is returned to the previous position 1 in the *same* step with probability 1. The transition diagram is shown in Figure 3.2.

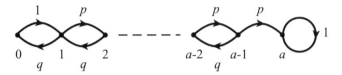

Fig. 3.2 *Transition diagram for the generous gambler problem.*

3.2 Unrestricted random walks

A simple random walk on a line or in one dimension occurs when a step forward (+1) has probability p and a step back (−1) has probability $q(= 1 - p)$. At the ith step, the modified Bernoulli random variable W_i is observed, and the position of the walk at the nth step is

$$X_n = X_0 + \sum_{i=1}^{n} W_i = X_{n-1} + W_n. \tag{3.1}$$

The random variable W_i can take the values +1 or −1, which is a *modified* Bernoulli random variable (see Section 1.7) in that the outcomes are $\{-1, +1\}$ rather than $\{0, 1\}$. In the gambler's ruin problem $X_0 = k$, but in the following discussion it is assumed, without loss of generality, that walks start from the origin so that $X_0 = 0$.

Several results have been derived for random walks that are restricted by boundaries. We now consider random walks without boundaries, or unrestricted random walks as they are known. In particular, we are interested in the position of the walk after a number of steps and the probability of a return to the origin, the start of the walk. As seen from equation (3.1) the position of the walk at step X_n simply depends on the position at the $(n - 1)$th step. This means that the simple random walk possesses what is known as the *Markov* property: the current state of the walk depends on its immediate previous state, not on the history of the walk up to the present state. Furthermore, $X_n = X_{n-1} \pm 1$, and we know that the transition probabilities from one position to another,

$$\mathbf{P}(X_n = j | X_{n-1} = j - 1) = p,$$

and
$$P(X_n = j | X_{n-1} = j + 1) = q,$$
are independent of n, the number of plays in the game or steps in the walk.

It is straightforward to find the mean and variance of X_n:

$$\mathbf{E}(X_n) = \mathbf{E}\left(\sum_{i=1}^{n} W_i\right) = n\mathbf{E}(W),$$

$$\mathbf{V}(X_n) = \mathbf{V}\left(\sum_{i=1}^{n} W_i\right) = n\mathbf{V}(W),$$

since the W_i are independent and identically distributed (iid) random variables and where W is the common or typical Bernoulli random variable in the sequence $\{W_i\}$. Thus

$$\mathbf{E}(W) = 1.p + (-1).q = p - q.$$

Since

$$\mathbf{V}(W) = \mathbf{E}(W^2) - [\mathbf{E}(W)]^2,$$

and

$$\mathbf{E}(W^2) = 1^2.p + (-1)^2 q = p + q = 1,$$

then

$$\mathbf{V}(W) = 1 - (p - q)^2 = 4pq.$$

Hence, the probability distribution of the position of the random walk at stage n has mean and variance

$$\mathbf{E}(X_n) = n(p - q), \quad \text{and} \quad \mathbf{V}(X_n) = 4npq.$$

If $p > \frac{1}{2}$ then we would correctly expect a drift away from the origin in a positive direction, and if $p < \frac{1}{2}$, it would be expected that the drift would be in the negative direction. However, since $\mathbf{V}(X_n)$ is proportional to n, and thus grows with increasing n, we would be increasingly uncertain about the position of the walker as n increases.

For the symmetric random walk, when $p = \frac{1}{2}$, the expected position after n steps is the origin. However, this is precisely the value of p that yields the maximum value of $\mathbf{V}(X_n) = 4npq = 4np(1-p)$ (check where $d\mathbf{V}(X_n)/dp = 0$). Thus, $\max_p \mathbf{V}(X_n) = n$.

Knowing the mean and standard variation of a random variable does not enable us to identify the probability distribution. For large n, however, we may apply the central limit theorem (see Grimmett and Welsh, 1986) which gives

$$Z_n = \frac{X_n - n(p - q)}{\sqrt{4npq}} \approx N(0, 1), \tag{3.2}$$

where $N(0, 1)$ is the normal distribution with mean 0 and variance 1. By applying a continuity correction, approximate probabilities may be obtained for the position of the walk. For example, consider the unrestricted random walk with $n = 100$ and $p = 0.7$. Then

$$\mathbf{E}(X_{100}) = 40, \qquad \mathbf{V}(X_{100}) = 84.$$

Suppose that we are interested in deriving the probability that the position of the walk at the 100th step is between 35 and 45 paces from the origin. Since X_n is discrete, we use the approximation

$$\mathbf{P}(35 \leq X_{100} \leq 45) \approx \mathbf{P}(34.5 < X_{100} < 45.5). \tag{3.3}$$

From the approximation (3.2), the inequality for X_{100} in the probability on the right in (3.3) is equivalent to

$$-0.60 = -\frac{5.5}{\sqrt{84}} < Z_{100} = \frac{X_{100} - 40}{\sqrt{84}} < \frac{5.5}{\sqrt{84}} = 0.60.$$

Hence

$$\mathbf{P}(-0.60 < Z_{100} < 0.60) = \Phi(0.60) - \Phi(-0.60) = 0.45,$$

where $\Phi(z)$ is the standard normal distribution function.

3.3 The probability distribution after n steps

As before, we assume that the walk starts at $x = 0$ with steps to the right or left occurring with probabilities p and $q = 1 - p$. The probability distribution of the random variable X_n, the position after n steps, is a more difficult problem. The position X_n, after n steps, can be written as

$$X_n = R_n - L_n,$$

where R_n is the number of right (positive) steps $(+1)$ and L_n is the number of left (negative) steps (-1). Furthermore

$$n = R_n + L_n.$$

Hence

$$R_n = \frac{1}{2}(n + X_n),$$

which is only an integer when n and X_n are both even or both odd (for example, to go from $x = 0$ to $x = 9$ must take an odd number of steps). Now, let $v_{n,x}$ be the probability that the walk is at state x after n steps. Assume that x is a positive integer. Thus

$$v_{n,x} = \mathbf{P}(X_n = x) = \mathbf{P}(R_n = \frac{1}{2}(n + x)),$$

and R_n is a binomial random variable with index n and probability p since the walk either moves to the right or not at every step, and the steps are independent. It follows that

$$v_{n,x} = \binom{n}{\frac{1}{2}(n+x)} p^{\frac{1}{2}(n+x)} q^{\frac{1}{2}(n-x)}, \tag{3.4}$$

where (n, x) are both even or both odd, $-n \le x \le n$. A similar argument can be constructed if x is a negative integer.

An alternative combinatorial argument for the case in which $p = \frac{1}{2}$ goes as follows. Starting from the origin, there are 2^n different paths of length n since there is a choice of right or left at each step. How many of these paths end at x (assumed positive)? As we saw above, the number of steps in the right direction must be $\frac{1}{2}(n+x)$, and the total number of paths must be the number of ways in which $\frac{1}{2}(n+x)$ can be chosen from n: this is

$$N_{n,x} = \binom{n}{\frac{1}{2}(n+x)}$$

provided $\frac{1}{2}(n+x)$ is an integer. Hence, by counting, the probability that the walk ends at x after n steps is given by the ratio of this number and the total number of paths since all paths are equally likely. Hence

$$v_{n,x} = \frac{N_{n,x}}{2^n} = \binom{n}{\frac{1}{2}(n+x)} \frac{1}{2^n}. \tag{3.5}$$

Example 3.1. Find the probability that the event $X_5 = 3$ occurs in a random walk with $p = 0.6$.

The event $X_5 = 3$ must occur as a result of observing four 1s and one -1 in any order. The probability of any one these sequences is $p^4 q$ and there are $\binom{5}{1}$ ways in which they can occur giving

$$\mathbf{P}(X_3 = 3) = \binom{5}{1} p^4 q = 0.259.$$

Alternatively, we can use the formula (3.4) which appears different at first sight but is the same since

$$\binom{5}{4} = \binom{5}{1}.$$

$\square$

The probability $v_{n,x}$ is the probability that the walk ends at state x after n steps: the walk could have overshot x before returning there. A related probability is the probability that the *first* visit to position x occurs at the nth step. This is sometimes known also as the *first passage* through x, and will be considered in the next section for $x = 0$. The following is a descriptive derivation of the associated probability generating function for the symmetric random walk in which the walk starts at the origin, and we consider the probability that it returns to the origin.

From the previous section, the probability that a walk is at the origin at step n is

$$v_{n,0} = \frac{1}{2^n}\binom{n}{\frac{1}{2}n} = p_n \quad \text{(say)}, \quad (n = 2, 4, 6, \ldots). \tag{3.6}$$

We assume that $p_n = 0$ if n is odd. Form a generating function:

$$H(s) = \sum_{n=0}^{\infty} p_n s^n = \sum_{n=0}^{\infty} p_{2n} s^{2n} = \sum_{n=0}^{\infty} \frac{1}{2^{2n}}\binom{2n}{n} s^{2n},$$

noting that $p_0 = 1$. Note that $H(s)$ is not a *probability* generating function since clearly $H(1) \neq 1$, which is a requirement of a probability generating function: subject to convergence, a generating function can be defined for any sequence. The binomial coefficient can be re-arranged as follows:

$$\binom{2n}{n} = \frac{(2n)!}{n!n!} = \frac{2n(2n-1)(2n-2)\ldots 3.2.1}{n!n!}$$
$$= \frac{2^n n!(2n-1)(2n-3)\ldots 3.1}{n!n!}$$
$$= \frac{2^{2n}}{n!}\frac{1}{2}\cdot\frac{3}{2}\ldots(n-\frac{1}{2}) = (-1)^n\binom{-\frac{1}{2}}{n}2^{2n}.$$

Hence

$$H(s) = \sum_{n=0}^{\infty}\binom{-\frac{1}{2}}{n}s^{2n} = (1-s^2)^{-\frac{1}{2}} \tag{3.7}$$

by the binomial theorem, provided $|s| < 1$. Note that this expansion guarantees that $p_n = 0$ if n is odd.

The distribution $\{p_n\}$ has some unusual features. The implication from (3.7) is that

$$\sum_{n=0}^{\infty} p_n = \lim_{s\to 1-} H(s) = \infty,$$

in other words the probabilities do *not* sum to one. (The notation $\lim_{s\to 1-}$ means that s approaches 1 through values of s that are less than 1.) This is a so-called *defective distribution* which still, however, gives the probability that the walk is at the origin at step n. We can estimate the behaviour of p_n for large n by using *Stirling's formula*, which is an asymptotic estimate for $n!$ for large n. The approximation is

$$n! \approx \sqrt{(2\pi)}n^{n+\frac{1}{2}}e^{-n}.$$

It follows from (3.6) that

$$p_{2n} = \frac{1}{2^{2n}}\binom{2n}{n} = \frac{1}{2^{2n}}\frac{(2n)!}{n!n!} \approx \frac{1}{2^{2n}}\frac{\sqrt{(2\pi)}(2n)^{2n+\frac{1}{2}}e^{-2n}}{[\sqrt{(2\pi)}(n)^{n+\frac{1}{2}}e^{-n}]^2} = \frac{1}{\sqrt{(\pi n)}}$$

for large n. Hence

$$np_n \to \infty,$$

which confirms that the series $\sum_{n=0}^{\infty} p_n$ must diverge.

3.4   First returns of the symmetric random walk

As we remarked in the previous section, a related probability is the one for the event in which the *first* visit to position x occurs at the nth step given that the walk starts at the origin; that is, the first passage through x. We shall look in detail at the case $x = 0$, which will lead to the probability of the first return to the origin. We shall approach the first passage by using total probability (Section 1.3). Let A be the event that $X_n = 0$, and let B_k be the event that the first visit to the origin occurs at the kth step. Then by the law of total probability

$$\mathbf{P}(A) = \sum_{k=1}^{n} \mathbf{P}(A|B_k)\mathbf{P}(B_k). \tag{3.8}$$

Let $f_k = \mathbf{P}(B_k)$. The conditional probability $\mathbf{P}(A|B_k)$ is the probability that the walk returns to the origin after $n - k$ steps, that is, $\mathbf{P}(A|B_k) = p_{n-k}$. As in the previous section, p_n is the probability that the walk is at the origin at step n. Note that $p_n = 0$ if n is odd. Hence, equation (3.8) can be written as

$$p_n = \sum_{k=1}^{n} p_{n-k} f_k. \tag{3.9}$$

Multiply both sides of (3.9) by s^n, and sum for $n \geq 1$ so that

$$H(s) - 1 = \sum_{n=1}^{\infty} p_n s^n = \sum_{n=1}^{\infty} \sum_{k=1}^{n} p_{n-k} f_k s^n. \tag{3.10}$$

Since $p_0 = 1$ and $f_0 = 0$, we can replace (3.10) by

$$H(s) - 1 = \sum_{n=1}^{\infty} p_n s^n = \sum_{n=0}^{\infty} \sum_{k=0}^{n} p_{n-k} f_k s^n = \sum_{n=0}^{\infty} p_n s^n \sum_{k=0}^{\infty} f_k s^k \tag{3.11}$$

the formula for the product of two power series. This also known as a *convolution* of the distributions. If

$$Q(s) = \sum_{n=0}^{\infty} f_n s^n$$

is the probability generating function of the first return distribution, then

$$H(s) - 1 = H(s)Q(s). \tag{3.12}$$

From (3.7) it follows that

$$Q(s) = [H(s) - 1]/H(s) = 1 - (1 - s^2)^{\frac{1}{2}}.$$

The probability that the walk will, at some step, return to the origin is

$$\sum_{n=1}^{\infty} f_n = Q(1) = 1;$$

in other words return is certain. In this walk the origin (or any point by translation) is *persistent*. However, the mean number of steps until this return occurs is

$$\sum_{n=1}^{\infty} n f_n = \lim_{s \to 1-} Q'(s) = \lim_{s \to 1-} \frac{s}{(1 - s^2)^{\frac{1}{2}}} = \infty.$$

In other words, a symmetric random walk which starts at the origin is certain to return there in the future, but, on average, it will take an infinite number of steps.

Example 3.2. Find the probability that a symmetric random walk starting from the origin returns there for the first time after 6 steps.

We require the coefficient of s^6 in the power series expansion of the pgf $Q(s)$, which is

$$\begin{aligned} Q(s) &= 1 - (1 - s^2)^{\frac{1}{2}} \\ &= 1 - [1 - \frac{1}{2}s^2 - \frac{1}{8}s^4 - \frac{1}{16}s^6 + O(s^8)] \\ &= \frac{1}{2}s^2 + \frac{1}{8}s^4 + \frac{1}{16}s^6 + O(s^8). \end{aligned}$$

Hence the probability of a first return at step 6 is $\frac{1}{16}$. $\qquad\qquad\square$

Long-term random walks are easy to compute. Figure 3.3 shows a computer simulation of two sample walks of 1000 steps starting at $k = 0$ with forward or backward steps equally likely. It might be expected intuitively that the walk would tend to oscillate about the starting position $k = 0$ by some law of averages. However, a feature of such walks is how few times the walk recrosses the axis $k = 0$. In fact, in the first case, after a brief oscillation about $k = 0$, the walk never returns to the start and finishes some 70 paces away. Intuition can be misleading in these problems. Remember the expected state is the average of many walks. A full discussion of this phenomena can be found in Feller (1968, Ch. 3).

3.5 Other random walks

We may extend the simple random walk in several ways. We may observe a discrete or continuous random variable over continuous time, or look at walks in two or more dimensions. This latter will not be considered theoretically here but it is easy to simulate.

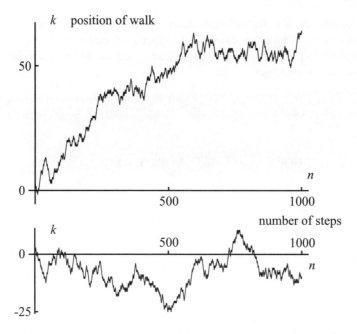

Fig. 3.3 *Two computer simulations of simple random walks of 1000 steps each.*

In later chapters we shall look at random walks with discrete steps that have a Poisson distribution over continuous time. We shall be exploring applications to birth and death processes where the outcome of interest is the number of individuals in a population at time t denoted by X_t. The population is the position of the random walk at time t. A homogeneous Poisson process of intensity α is such that, for any two times $t_1, t_2, (t_1 < t_2)$, the random variable

$$X_{t_2} - X_{t_1}, \quad (X_{t_2} \geq X_{t_1}),$$

which is the step length or the number of events between the two times, has a Poisson distribution with parameters $\alpha(t_2 - t_1)$. Also, for any overlapping time intervals, the step lengths are independent Poisson random variables. This implies that if $X_0 = 0$, that is, the walk starts from the origin, then X_t has a Poisson distribution with parameter λt. Hence

$$\mathbf{E}(X_t) = \mathbf{V}(X_t) = \lambda t.$$

If the walk starts from a point c, say, then

$$\mathbf{E}(X_t) = c + \lambda t, \quad \text{and} \quad \mathbf{V}(X_t) = \lambda t.$$

It will be shown later that the time interval between successive events has an exponential distribution with parameter λ.

Let us now return to the simple symmetric unrestricted random walk. Suppose that it is observed over a time interval $(0, t]$ at equally spaced times $0, d, 2d, \ldots, t$ and that the step lengths are (for convenience) $\sqrt{d}$ so that at the mth step the position W_{md} is such that

$$P(W_{md} = \sqrt{d}) = P(W_{md} = -\sqrt{d}) = \frac{1}{2}.$$

Then

$$E(W_{md}) = 0, \quad \text{and} \quad V(W_{md}) = E(W_{md}^2) = d.$$

Now, as before, assume that $X_0 = 0$. Then

$$E(X_t) = 0, \quad \text{and} \quad V(X_t) = t,$$

since W_{md} are independent and identically distributed modified Bernoulli random variables. This is a similar result to that obtained earlier where $E(X_n) = 0$ and $V(X_n) = n$. If d is allowed to approach zero, then this has the effect of approximating to a *continuous* time process in which there are small step lengths. Also, the number of individuals being observed will be large. The central limit theorem may now be applied to derive the asymptotic distribution of X_t, which is normal with mean 0 and variance t. In the limit, the random variable X_t for all t forms what is known as a *Wiener process* or *Brownian motion*.

In general, X_t will form a Wiener process if, for any two times t_1 and t_2 ($t_1 < t_2$),

$$X_{t_2} - X_{t_1}$$

has a normal distribution with mean 0 and variance $t_2 + t_1$, and these step lengths for any non-overlapping time intevals are independent normally distributed random variables.

Problems

3.1. In a simple random walk, the probability that the walk advances by one step is p and retreats by one step is $q = 1 - p$. At step n let the position of the walker be the random variable X_n. If the walk starts at $x = 0$, enumerate all possible sample paths that lead to the value $X_4 = -2$. Verify that

$$P\{X_4 = -2\} = \binom{4}{1} pq^3.$$

3.2. A symmetric random walk starts from the origin. Find the probability that the walker is at the origin at step 8. What is the probability that the walker is at the origin but that it is not the first visit there?

3.3. An asymmetric walk with parameter p starts at the origin. From equation (3.4), the probability that the walk reaches x in n steps is given by

$$v_{n,x} = \binom{n}{\frac{1}{2}(n+x)} p^{\frac{1}{2}(n+x)} q^{\frac{1}{2}(n-x)},$$

where n and x are both even or both odd. If $n = 4$, show that the mean value position x is $4(p - q)$, confirming the result in Section 3.2.

3.4. The pgf for the first return distribution $\{f_n\}$ to the origin in a symmetric random walk is given by

$$Q(s) = 1 - (1 - s^2)^{\frac{1}{2}},$$

(see Section 3.4).

(a) Using the binomial theorem find a formula for f_n, the probability that the first return occurs at the nth step.
(b) What is the variance of the probability distribution $f_n, n = 1, 2, \ldots$?

3.5. A coin is spun n times and the sequence of heads and tails is recorded. What is the probability that the number of heads equals the number of tails after n spins?

3.6. For an asymmetric walk with parameters p and $q = 1 - p$, the probability that the walk is at the origin after n steps is

$$q_n = v_{n,0} = \begin{cases} \binom{n}{\frac{1}{2}n} p^{\frac{1}{2}n} q^{\frac{1}{2}n} & n \text{ even}, \\ 0 & n \text{ odd} \end{cases}$$

from equation (3.4). Show that its generating function is

$$P(s) = (1 - 4pqs^2)^{-\frac{1}{2}}.$$

If $p \neq q$, show that the mean number of steps to the return is

$$m = P'(1) = \frac{4pqs}{(1 - 4pqs^2)^{\frac{3}{2}}}\bigg|_{s=1} = \frac{4pq}{(1 - 4pq)^{\frac{3}{2}}}.$$

What is its variance?

3.7. Using the results of Problem 3.6 and equation (3.12) relating to the generating functions of the returns and first returns to the origin, namely

$$H(s) - 1 = H(s)Q(s),$$

which is still valid for the asymmetric walk, show that

$$Q(s) = 1 - (1 - 4pqs^2)^{\frac{1}{2}},$$

where $p \neq q$. Show that a first return to the origin is not certain, unlike the situation in the symmetric walk. Find the mean number of steps to the first return.

3.8. A symmetric random walk starts from the origin. Show that the walk does not revisit the origin in the first $2n$ steps with probability

$$h_n = 1 - f_2 - f_4 - \cdots - f_{2n},$$

where f_n is the probability that a first return occurs at the nth step.
   The generating function for the sequence $\{f_n\}$ is

$$Q(s) = 1 - (1 - s^2)^{\frac{1}{2}},$$

(see Section 3.4). Show that

$$f_n = \begin{cases} (-1)^{n+1}\binom{\frac{1}{2}}{n} & n \text{ even} \\ 0 & n \text{ odd} \end{cases} \qquad (n = 1, 2, 3, \ldots).$$

Show that h_n satisfies the first-order difference equation

$$h_{n+1} - h_n = (-1)^{n+1}\binom{\frac{1}{2}}{n+1}.$$

Verify that this equation has the general solution

$$h_n = C + \binom{2n}{n}\frac{1}{2^{2n}},$$

where C is a constant. By calculating h_1, confirm that the probability of no return to the origin in the first $2n$ steps is $\binom{2n}{n}/2^{2n}$.

3.9. A walk can be represented as a connected graph between coordinates (n, y) where the ordinate y is the position on the walk, and the abscissa n represents the number of steps. A walk of 7 steps which joins $(0, 1)$ and $(7, 2)$ is shown in Figure 3.4. Suppose that a walk starts at $(0, y_1)$ and finishes at (n, y_2), where $y_1 > 0$, $y_2 > 0$ and $n + y_2 - y_1$ is an even number. Suppose also that the walk first visits the origin at $n = n_1$. Reflect that part of the path for which $n \leq n_1$ in the n-axis (see Figure 3.4), and use a reflection argument to show that the number of paths from $(0, y_1)$ to (n, y_2) which touch or cross the n-axis equals the number of *all* paths from $(0, -y_1)$ to (n, y_2). This is known as the *reflection principle*.

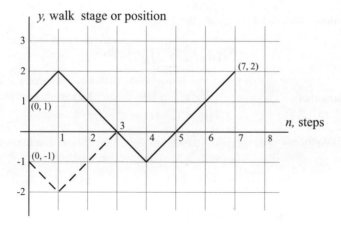

Fig. 3.4 *Representation of a random walk.*

3.10. A walk starts at $(0, 1)$ and returns to $(2n, 1)$ after $2n$ steps. Using the reflection principle (see Problem 3.9) show that there are

$$\frac{(2n)!}{n!(n+1)!}$$

different paths between the two points that never revisit the origin. What is the probability that the walk ends at $(2n, 1)$ after $2n$ steps without ever visiting the origin, assuming that the random walk is symmetric?

Show that the probability that the first visit to the origin after $2n+1$ steps is

$$p_n = \frac{1}{2^{2n+1}} \frac{(2n)!}{n!(n+1)!}.$$

For large n, apply Stirling's approximate formula $n! \approx \sqrt{(2\pi)} n^{n+\frac{1}{2}} e^{-n}$ to the factorials in this probability. Show that

$$n p_n \approx \frac{1}{2\sqrt{(n\pi)}(1+\frac{1}{n})^{3/2}}.$$

(You will need the approximation $(1+\frac{1}{n})^n \approx e$ for large n also.) Hence deduce that the mean number of steps to the first visit to the origin must be infinite.

3.11. A symmetric random walks starts at the origin. Let $f_{n,1}$ be the probability that the first visit to position $x = 1$ occurs at the nth step. Obviously, $f_{2n,1} = 0$. The result from Problem 3.10 can be adapted to give

$$f_{2n+1,1} = \frac{1}{2^{2n+1}} \frac{(2n)!}{n!(n+1)!}, \qquad (n = 0, 1, 2, \ldots).$$

Suppose that its pgf is

$$G_1(s) = \sum_{n=0}^{\infty} f_{2n+1,1} s^{2n+1}.$$

Show that

$$G_1(s) = [1 - (1 - s^2)^{\frac{1}{2}}]/s.$$

[Hint: the identity

$$\frac{1}{2^{2n+1}} \frac{(2n)!}{n!(n+1)!} = (-1)^n \binom{\frac{1}{2}}{n+1}, \qquad (n = 0, 1, 2, \ldots)$$

is useful in the derivation of $G_1(s)$.]

Show that any walk starting at the origin is certain to visit $x > 0$ at some future step, but that the mean number of steps in achieving this is infinite.

3.12. A symmetric random walk starts at the origin. Let $f_{n,x}$ be the probability that the first visit to position x occurs at the nth step (as usual, $f_{n,x} = 0$ if $n + x$ is an odd number). Explain why

$$f_{n,x} = \sum_{k=1}^{n-1} f_{n-k,x-1} f_{k,1}, \qquad (n \ge x > 1).$$

If $G_x(s)$ is its pgf, deduce that

$$G_x(s) = \{G_1(s)\}^x,$$

where $G_1(s)$ is given explicitly in Problem 3.11. What are the probabilities that the walk first visits $x = 3$ at the steps $n = 3$, $n = 5$ and $n = 7$?

3.13. Problem 3.12 looks at the probability of a first visit to position $x \ge 1$ at the nth step in a symmetric random walk that starts at the origin. Why is the probability generating function for the first visit to position x where $|x| \ge 1$ given by

$$G_x(s) = \{G_1(s)\}^{|x|},$$

where $G_1(s)$ is defined in Problem 3.11?

3.14. An asymmetric walk has parameters p and $q = 1 - p \ne p$. Let $g_{n,1}$ be the probability that the first visit to $x = 1$ occurs at the nth step. As in Problem 3.11, $g_{2n,1} = 0$. It was effectively shown in Problem 3.10 that the number of paths from the origin which return to the origin after $2n$ steps is

$$\frac{(2n)!}{n!(n+1)!}.$$

Explain why

$$g_{2n+1,1} = \frac{(2n)!}{n!(n+1)!} p^{n+1} q^n.$$

Suppose that its pgf is

$$G_1(s) = \sum_{n=0}^{\infty} g_{2n+1,1} s^{2n+1}.$$

Show that

$$G_1(s) = [1 - (1 - 4pqs^2)^{\frac{1}{2}}]/(2qs).$$

(The identity in Problem 3.11 is required again.)

What is the probability that the walk ever visits $x > 0$? How does this result compare with that for the symmetic random walk?

What is the pgf for the distribution of first visits of the walk to $x = -1$ at step $2n + 1$?

3.15. It was shown in Section 3.3 that, in a random walk with parameters p and $q = 1 - p$, the probability that a walk is at position x at step n is given by

$$v_{n,x} = \binom{n}{\frac{1}{2}(n+x)} p^{\frac{1}{2}(n+x)} q^{\frac{1}{2}(n-x)}, \qquad |x| \le n,$$

where $\frac{1}{2}(n + x)$ must be an integer. Verify that $v_{n,x}$ satisfies the difference equation

$$v_{n+1,x} = pv_{n,x-1} + qv_{n,x+1},$$

subject to the initial conditions

$$v_{0,0} = 1, \qquad v_{n,x} = 0, \quad (x \ne 0).$$

Note that this difference equation has differences on two arguments.

Can you develop a direct argument that justifies the difference equation for the random walk?

3.16. In the usual notation, $v_{2n,0}$ is the probability that, in a symmetric random walk, the walk visits the origin after $2n$ steps. Using the difference equation from Problem 3.15, $v_{2n,0}$ satisfies

$$v_{2n,0} = \frac{1}{2} v_{2n-1,-1} + \frac{1}{2} v_{2n-1,1} = v_{2n-1,1}.$$

How can the last step be justified? Let

$$G_1(s) = \sum_{n=1}^{\infty} v_{2n-1,1} s^{2n-1}$$

be the pgf of the distribution $\{v_{2n-1,1}\}$. Show that

$$G_1(s) = [H(s) - 1]/s,$$

where the formula for $H(s)$ is given by

$$H(s) = \sum_{n=0}^{\infty} v_{n,0} s^n = (1 - s^2)^{-\frac{1}{2}}$$

(see equation (3.7)). By expanding $G_1(s)$ as a power series in s show that

$$v_{2n-1,1} = \binom{2n-1}{n} \frac{1}{2^{2n-1}}.$$

By a repetition of the argument show that

$$G_2(s) = \sum_{n=0}^{\infty} v_{2n,2} s^{2n} = [(2-s)(1-s^2)^{-\frac{1}{2}} - 2]/s.$$

3.17. A random walk takes place on a circle that is marked out with N positions. Thus, as shown in Figure 3.5, position N is the same as position O. This is known as the *cyclic random walk* of *period N*. A symmetric random walk, starts at O. What is the probability that the walk is at O after n steps in the cases:

(a) $n < N$;
(b) $N \le n < 2N$?

Distinguish carefully the cases in which n and N are even and odd.

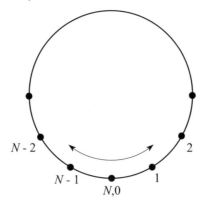

Fig. 3.5 *The cyclic random walk of period N.*

3.18. An unrestricted random walk with parameters p and q starts from the origin, and lasts for 50 paces. Estimate the probability that the walk ends at 12 or more paces from the origin in the cases:

(a) $p = q = \frac{1}{2}$;
(b) $p = 0.6, q = 0.4$.

3.19. In an unrestricted random walk, with parameters p and q, for what value of p are the mean and variance of the probability distribution of the position of the walk at stage n the same?

3.20. Two walkers each perform symmetric random walks with synchronized steps, both starting from the origin at the same time. What is the probability that they are both at the origin at step n?

3.21. A random walk takes place on a two-dimensional lattice as shown in Figure 3.6. In the example shown the walk starts at $(0, 0)$ and ends at $(2, -1)$ after 13 steps. In this walk direct diagonal steps are not permitted. We

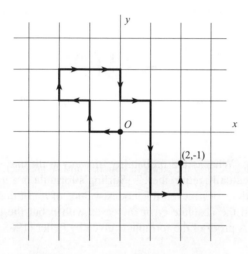

Fig. 3.6 *A two-dimensional random walk.*

are interested in the probability that the symmetric random walk, which starts at the origin, has returned there after $2n$ steps. Symmetry in the two-dimensional walk means that there is a probability of $\frac{1}{4}$ that, at any position, the walk goes right, left, up, or down at the next step. The total number of different walks of length $2n$, which start at the origin, is 4^{2n}. For the walk considered, the number of right steps (positive x direction) must equal the number of left steps, and the number of steps up (positive y direction) must equal those down. In addition, the number of right steps must range from 0 to n, and the corresponding steps up from n to 0. Explain why the probability that the walk returns to the origin after $2n$ steps is

$$p_{2n} = \frac{(2n)!}{4^{2n}} \sum_{r=0}^{n} \frac{1}{[r!(n-r)!]^2}.$$

Prove the two identities

$$\frac{(2n)!}{[r!(n-r)!]^2} = \binom{2n}{n}\binom{n}{r}^2, \qquad \binom{2n}{n} = \sum_{r=0}^{n}\binom{n}{r}^2.$$

[Hint: compare the coefficients of x^n in $(1+x)^{2n}$ and $[(1+x)^n]^2$.] Hence show that

$$p_{2n} = \frac{1}{4^{2n}}\binom{2n}{n}^2.$$

Calculate p_2, p_4, $1/(\pi p_{40})$, $1/(\pi p_{80})$. How do you would you guess that p_{2n} behaves for large n?

3.22. It was shown in Problem 3.21 that the probability p_{2n} of a return to the origin after $2n$ steps in a symmetric two-dimensional random walk is given by

$$p_{2n} = \frac{1}{4^{2n}}\binom{2n}{n}^2.$$

Apply Stirling's approximation $n! \approx \sqrt{(2\pi)}n^{n+\frac{1}{2}}e^{-n}$ to p_{2n} for large n. Show that

$$p_{2n} \approx \frac{1}{\pi n}$$

for large n. What is the probability that any walk that starts at the origin will eventually return there? [Stirling's formula is a useful *asymptotic* approximation for $n!$, even for modest values of n. In this asymptotic approximation the absolute error increases with n but the percentage error decreases. A proof of the formula is given by Feller (1968, Section 2.9).]

4
Markov chains

4.1 States and transitions

The random walk discussed in the previous chapter is a special case of a more general *Markov process*. Suppose that a random process goes through a discrete sequence of steps or trials numbered $n = 0, 1, 2, \ldots$, where the outcome of the nth trial is the random variable X_n. This discrete random variable can take one of the values $i = 1, 2, \ldots$. The actual outcomes are called the *states* of the system, and are denoted by E_i $(i = 1, 2, \ldots)$. In most, but not quite all, cases in this text, we shall investigate systems with a finite number, m, of states $E_1, E_2, \ldots E_m$. However, in the unrestricted random walk of Section 3.2, for example, the states will be all possible locations of the walker, which must be infinite in number. Since both forward and backward steps can occur, the list of states can be expressed as the unbounded sequence,

$$\{\ldots E_{-2}, E_{-1}, E_0, E_2, \ldots\}.$$

In this sequence, E_r is the state in which the walker is at step n; that is, the state in which $X_n = r$.

In general, a *chain* is a discrete-time process in which the random variable X_n undergoes a sequence of changes at a sequence of times or steps. The times or steps can be variable in duration. If the random variables X_{n-1} and X_n take the values $X_{n-1} = i$ and $X_n = j$, then the system has made a *transition* $E_i \rightarrow E_j$, that is, a transition from state E_i to state E_j at the nth trial. Note that i can equal j, so that transitions within the same state may be possible. We need to assign probabilities to the transitions $E_i \rightarrow E_j$. Generally in chains, the probability that $X_n = j$ will depend on the whole sequence of random variables starting with the initial value X_0. The *Markov chain* has the characteristic property that the probability that $X_n = j$ depends *only* on the immediate previous state of the system. Formally, this means that we need no further information at each step other than, for each i and j,

$$\mathbf{P}\{X_n = j | X_{n-1} = i\},$$

which means the probability that $X_n = j$ given that $X_{n-1} = i$: this probability is *independent* of the values of $X_{n-2}, X_{n-3}, \ldots, X_0$.

In some chains the probabilities $\mathbf{P}\{X_n = j | X_{n-1} = i\}$ are functions of n, the step or trial number. If this is not the case, so that the probabilities are the same at every step, then the chain is said to be *homogeneous*.

4.2 Transition probabilities

For a finite homogeneous Markov chain with m states $E_1, E_2, \ldots, E_m$, introduce the notation

$$p_{ij} = \mathbf{P}\{X_n = j | X_{n-1} = i\}, \tag{4.1}$$

where $i, j = 1, 2, \ldots, m$. If $p_{ij} > 0$, then we say that state E_i can *communicate* with E_j: two-way communication is possible if, additionally, $p_{ji} > 0$. Obviously for each fixed i, the list $\{p_{ij}\}$ is a *probability distribution*, since at any step one of the outcomes $E_1, E_2, \ldots, E_m$ must occur: the states E_i, $(i = 1, 2, \ldots m)$ are exhaustive and mutually exclusive. The numbers p_{ij} are known as the *transition probabilities* of the chain, and must satisfy

$$p_{ij} \geq 0, \quad \sum_{j=1}^{m} p_{ij} = 1$$

for each $i = 1, 2, \ldots, m$.

Transition probabilities form an $m \times m$ array that can be combined into a *transition matrix* T, where

$$T = [p_{ij}] = \begin{bmatrix} p_{11} & p_{12} & \cdots & p_{1m} \\ p_{21} & p_{22} & \cdots & p_{2m} \\ \vdots & \vdots & \ddots & \vdots \\ p_{m1} & p_{m2} & \cdots & p_{mm} \end{bmatrix} \tag{4.2}$$

Note that each row of T is a probability distribution. Any square matrix for which $p_{ij} \geq 0$ and $\sum_{j=1}^{m} p_{ij} = 1$ is said to be *row-stochastic*

Example 4.1. The matrices $A = [a_{ij}]$ and $B = [b_{ij}]$ are $m \times m$ row-stochastic matrices. Show that $C = AB$ is also row-stochastic.

By the multiplication rule for matrices

$$C = AB = [a_{ij}][b_{ij}] = \left[\sum_{k=1}^{m} a_{ik} b_{kj} \right].$$

Hence c_{ij}, the general element of C, is given by

$$c_{ij} = \sum_{k=1}^{m} a_{ik} b_{kj}.$$

Since $a_{ij} \geq 0$ and $b_{ij} \geq 0$ for all $i, j = 1, 2, \ldots, m$, it follows that $c_{ij} \geq 0$. Also

$$\sum_{j=1}^{m} c_{ij} = \sum_{j=1}^{m} \sum_{k=1}^{m} a_{ik} b_{kj} = \sum_{k=1}^{m} a_{ik} \sum_{j=1}^{m} b_{kj} = \sum_{k=1}^{m} a_{ik} \cdot 1 = 1,$$

since $\sum_{j=1}^{m} b_{kj} = 1$ and $\sum_{k=1}^{m} a_{ik} = 1$. $\qquad \square$

It follows from this example that any power T^n of the transition matrix T must be row-stochastic.

The absolute probability $p_j^{(n)}$

One further probability that will be of interest is the probability of outcome E_j after n steps, given an *initial probability distribution* $\{p_i^{(0)}\}$. Here $p_i^{(0)}$ is the probability that, initially, the system occupies state E_i. Of course we must have $\sum_{i=1}^{m} p_i^{(0)} = 1$. Let $p_j^{(1)}$ be the probability E_j is occupied after one step. Then, by the law of total probability (see Section 1.3)

$$p_j^{(1)} = \sum_{i=1}^{m} p_i^{(0)} p_{ij}. \tag{4.3}$$

We can express this more conveniently in vector form. Let $\mathbf{p}^{(0)}$ and $\mathbf{p}^{(1)}$ be the *probability (row) vectors* given by

$$\mathbf{p}^{(0)} = \begin{bmatrix} p_1^{(0)} & p_2^{(0)} & \cdots & p_m^{(0)} \end{bmatrix} \tag{4.4}$$

and

$$\mathbf{p}^{(1)} = \begin{bmatrix} p_1^{(1)} & p_2^{(1)} & \cdots & p_m^{(1)} \end{bmatrix}. \tag{4.5}$$

Here, $\mathbf{p}^{(0)}$ is the initial distribution, and the components of $\mathbf{p}^{(1)}$ will be the probabilities that each of the states $E_1, E_2, \ldots, E_{(m)}$ is reached after one step. Equation (4.3) can be represented as a matrix product as follows:

$$\mathbf{p}^{(1)} = \begin{bmatrix} p_j^{(1)} \end{bmatrix} = \begin{bmatrix} \sum_{i=1}^{m} p_i^{(0)} p_{ij} \end{bmatrix} = \mathbf{p}^{(0)} T,$$

where T is the transition matrix given by (4.2). If $\mathbf{p}^{(2)}$ is the distribution after two steps, then

$$\mathbf{p}^{(2)} = \mathbf{p}^{(1)} T = \mathbf{p}^{(0)} TT = \mathbf{p}^{(0)} T^2.$$

Hence, after n steps

$$\mathbf{p}^{(n)} = \mathbf{p}^{(n-1)} T = \mathbf{p}^{(0)} T^n, \tag{4.6}$$

where

$$\mathbf{p}^{(n)} = \begin{bmatrix} p_1^{(n)} & p_2^{(n)} & \cdots & p_m^{(n)} \end{bmatrix}. \tag{4.7}$$

More generally

$$\mathbf{p}^{(n+r)} = \mathbf{p}^{(r)} T^n.$$

In (4.7), the component $p_j^{(n)}$ is the *absolute* or *unconditional probability of outcome* E_j at the nth step given by the initial distribution $\mathbf{p}^{(0)}$, that is, $\mathbf{P}\{X_n = j\} = p_j^{(n)}$. Note that

$$\sum_{j=1}^{m} p_j^{(n)} = 1.$$

The n-step transition probability $p_{ij}^{(n)}$

We now define $p_{ij}^{(n)}$ as the probability that the chain is in state E_j after n steps *given that the chain started in* state E_i. The first step transition probabilities $p_{ij}^{(1)} = p_{ij}$ are simply the elements of the transition matrix T. We intend to find a formula for $p_{ij}^{(n)}$. Now, by definition,

$$p_{ij}^{(n)} = \mathbf{P}(X_n = j | X_0 = i),$$

and by the Markov property of chains (Section 4.1)

$$p_{ij}^{(n)} = \sum_{k=1}^{m} \mathbf{P}(X_n = j, X_{n-1} = k | X_0 = i)$$

for $n \geq 2$, since the chain must have passed through one of all the m possible states at step $n - 1$.

For any three events A, B and C, we have available the identity

$$\mathbf{P}(A \cap B | C) = \mathbf{P}(A | B \cap C)\mathbf{P}(B | C),$$

(see Example 1.4). Interpreting A as $X_n = j$, B as $X_{n-1} = k$ and C as $X_0 = i$, it follows that

$$p_{ij}^{(n)} = \sum_{k=1}^{m} \mathbf{P}(X_n = j | X_{n-1} = k, X_0 = i)\mathbf{P}(X_{n-1} = k | X_0 = i)$$

$$= \sum_{k=1}^{m} \mathbf{P}(X_n = j | X_{n-1} = k)\mathbf{P}(X_{n-1} = k | X_0 = i)$$

$$= \sum_{k=1}^{m} p_{kj}^{(1)} p_{ik}^{(n-1)},$$

using the Markov property again. These are known as the *Chapman–Kolmogorov equations*. Putting n successively equal to 2, 3, ..., we find that the matrices with these elements are

$$\left[p_{ij}^{(2)} \right] = \left[p_{ik}^{(1)} p_{kj}^{(1)} \right] = T^2,$$

$$\left[p_{ij}^{(3)} \right] = \left[p_{ik}^{(2)} p_{kj}^{(1)} \right] = T^3,$$

since $p_{ik}^{(2)}$ are the elements of T^2, and so on. Hence

$$p_{ij}^{(n)} = T^n.$$

Example 4.2. In a certain region the weather patterns have the following sequence. A day is described as sunny *(S)* if the sun shines for more than 50% of daylight hours and cloudy *(C)* if the sun shines for less than 50% of daylight hours. Data indicate that if it is cloudy one day then it is equally likely to be cloudy or sunny on the next day; if it is sunny there is a probability $\frac{1}{3}$ that it is cloudy and $\frac{2}{3}$ that it is sunny the next day.

(i) Construct the transition matrix T for this process.
(ii) If it is cloudy today what are the probabilities that it is (a) cloudy, (b) sunny, in three days' time?
(iii) Compute T^5 and T^{10}. How do you think that T^n behaves as $n \to \infty$? How does $\mathbf{p}^{(n)}$ behave as $n \to \infty$? Do you expect the limit to depend on $\mathbf{p}^{(0)}$?

(i) It is assumed that the process is Markov and homogeneous so that transition probabilities depend only on the state of the weather on the previous day. This a two-state Markov chain with states

$$E_1 = (\text{weather cloudy, } C), \qquad E_2 = (\text{weather sunny, } S).$$

The transition probabilites can be represented by the table below, which defines the transition matrix T:

	C	S
C	$\frac{1}{2}$	$\frac{1}{2}$
S	$\frac{1}{3}$	$\frac{2}{3}$

or $T = \begin{bmatrix} \frac{1}{2} & \frac{1}{2} \\ \frac{1}{3} & \frac{2}{3} \end{bmatrix}$.

The actual transition probabilities are

$$p_{11} = \frac{1}{2}, \quad p_{12} = \frac{1}{2}, \quad p_{21} = \frac{1}{3}, \quad p_{22} = \frac{2}{3}.$$

(ii) Measuring steps from today, we define

$$\mathbf{p}^{(0)} = \left[\ p_1^{(0)} \ \ p_2^{(0)} \ \right] = \left[\ 1 \ \ 0 \ \right]$$

which means that it is cloudy today. In three days' time.

$$\mathbf{p}^{(3)} = \mathbf{p}^{(0)} T^3 = \left[\ 1 \ \ 0 \ \right] \begin{bmatrix} \frac{1}{2} & \frac{1}{2} \\ \frac{1}{3} & \frac{2}{3} \end{bmatrix}^3$$

$$= \left[\ 1 \ \ 0 \ \right] \begin{bmatrix} 29/72 & 43/72 \\ 43/108 & 65/108 \end{bmatrix}$$

$$= \left[\ 29/72 \ \ 43/72 \ \right] = \left[\ 0.403 \ \ 0.600 \ \right]$$

Hence the probabilities of cloudy or sunny weather in three days' time are respectively:

(a) $p_1^{(3)} = 29/72$

(b) $p_2^{(3)} = 43/72$.

(iii) The computed values of T^5 and T^{10} are (to 6 decimal places):

$$T^5 = \begin{bmatrix} 0.400077 & 0.599923 \\ 0.399949 & 0.600051 \end{bmatrix}, \qquad T^{10} = \begin{bmatrix} 0.400000 & 0.600000 \\ 0.400000 & 0.600000 \end{bmatrix}.$$

Powers of matrices can be easily computed using software such as *Mathematica*. It appears that

$$T^n \to \begin{bmatrix} 0.4 & 0.6 \\ 0.4 & 0.6 \end{bmatrix} = Q,$$

say, as $n \to \infty$. From (4.6)

$$\mathbf{p}^{(n)} = \mathbf{p}^{(0)} T^n.$$

If $T^n \to Q$ as $n \to \infty$, then we might expect

$$\mathbf{p}^{(n)} \to \mathbf{p}^{(0)} Q = \begin{bmatrix} p_1^{(0)} & p_2^{(0)} \end{bmatrix} \begin{bmatrix} 0.4 & 0.6 \\ 0.4 & 0.6 \end{bmatrix}$$

$$= \begin{bmatrix} (p_1^{(0)} + p_2^{(0)})0.4 & (p_1^{(0)} + p_2^{(0)})0.6 \end{bmatrix}$$

$$= \begin{bmatrix} 0.4 & 0.6 \end{bmatrix}$$

since $p_1^{(0)} + p_2^{(0)} = 1$. It appears that for all the transition matrices in the problem, $\lim_{n\to\infty} \mathbf{p}^n$ is independent of $\mathbf{p}^{(0)}$. The limit indicates that, in the long run, 40% of days are cloudy and 60% are sunny. ☐

This example indicates that it would be useful if we had a general algebraic formula for the nth power of a matrix. The algebra required for this aspect of Markov chains will be looked at in the next two sections.

4.3 General two-state Markov chains

Consider the two-state chain with transition matrix

$$T = \begin{bmatrix} 1 - \alpha & \alpha \\ \beta & 1 - \beta \end{bmatrix}, \qquad 0 < \alpha, \beta < 1.$$

We want to construct a formula for T^n. First, find the *eigenvalues* (λ) of T: they are given by the solutions of the determinant equation $\det(T - \lambda I_2) = 0$, where I_2 is the 2×2 identity matrix given by

$$I_2 = \begin{bmatrix} 1 & 0 \\ 0 & 1 \end{bmatrix}.$$

The *characteristic equation*, as it is known, is

$$\begin{vmatrix} 1 - \alpha - \lambda & \alpha \\ \beta & 1 - \beta - \lambda \end{vmatrix} = 0, \quad \text{or} \quad (1 - \alpha - \lambda)(1 - \beta - \lambda) - \alpha\beta = 0.$$

Hence λ satisfies the quadratic equation

$$\lambda^2 - \lambda(2 - \alpha - \beta) + 1 - \alpha - \beta = 0,$$

or

$$(\lambda - 1)(\lambda - 1 + \alpha + \beta) = 0. \tag{4.8}$$

Let the eigenvalues of T be $\lambda_1 = 1$ and $\lambda_2 = 1 - \alpha - \beta = r$, say. We now find the *eigenvectors* associated with each eigenvalue. Let $\mathbf{r}_1$ be the (column) eigenvector of λ_1 defined by

$$(T - \lambda_1 I_2)\mathbf{r}_1 = \mathbf{0}, \quad \text{or} \quad \begin{bmatrix} 1 - \alpha - \lambda_1 & \alpha \\ \beta & 1 - \beta - \lambda_1 \end{bmatrix} \mathbf{r}_1 = \mathbf{0},$$

or

$$\begin{bmatrix} -\alpha & \alpha \\ \beta & -\beta \end{bmatrix} \mathbf{r}_1 = \mathbf{0}.$$

Choose *any* (non-zero) solution of this equation. So let

$$\mathbf{r}_1 = \begin{bmatrix} 1 \\ 1 \end{bmatrix}.$$

Similarly, the second eigenvector $\mathbf{r}_2$ satisfies

$$\begin{bmatrix} 1 - \alpha - \lambda_2 & \alpha \\ \beta & 1 - \beta - \lambda_2 \end{bmatrix} \mathbf{r}_2 = \mathbf{0}, \quad \text{or} \quad \begin{bmatrix} \beta & \alpha \\ \beta & \alpha \end{bmatrix} \mathbf{r}_2 = \mathbf{0}.$$

In this case we can choose

$$\mathbf{r}_2 = \begin{bmatrix} -\alpha \\ \beta \end{bmatrix}$$

Now form the matrix C, which has the eigenvectors $\mathbf{r}_1$ and $\mathbf{r}_2$ as columns, so that

$$C = [\; \mathbf{r}_1 \quad \mathbf{r}_2 \;] = \begin{bmatrix} 1 & -\alpha \\ 1 & \beta \end{bmatrix}. \tag{4.9}$$

Now invert C:

$$C^{-1} = \frac{1}{\alpha + \beta} \begin{bmatrix} \beta & \alpha \\ -1 & 1 \end{bmatrix}.$$

If we now expand the matrix product $C^{-1}TC$, we now find that

$$\begin{aligned} C^{-1}TC &= \frac{1}{\alpha + \beta} \begin{bmatrix} \beta & \alpha \\ -1 & 1 \end{bmatrix} \begin{bmatrix} 1 - \alpha & \alpha \\ \beta & -\beta \end{bmatrix} \begin{bmatrix} 1 & -\alpha \\ 1 & \beta \end{bmatrix} \\ &= \frac{1}{\alpha + \beta} \begin{bmatrix} \beta & \alpha \\ -1 & 1 \end{bmatrix} \begin{bmatrix} 1 & -\alpha r \\ 1 & \beta r \end{bmatrix} \\ &= \begin{bmatrix} 1 & 0 \\ 0 & r \end{bmatrix} = \begin{bmatrix} \lambda_1 & 0 \\ 0 & \lambda_2 \end{bmatrix} = D, \end{aligned} \tag{4.10}$$

say. Now D is a *diagonal matrix* with the eigenvalues of T as its diagonal elements: this process is known in linear algebra as the *diagonalization of a matrix*. The result is significant since diagonal matrices are easy to multiply. From (4.10), if we premultiply by matrix C and post-multiply by C^{-1}, then we find that $T = CDC^{-1}$. Thus

$$T^2 = (CDC^{-1})(CDC^{-1}) = (CD)(C^{-1}C)(DC^{-1})$$
$$= (CD)I_2(DC^{-1}) = CDDC^{-1} = CD^2C^{-1},$$

where

$$D^2 = \begin{bmatrix} \lambda_1^2 & 0 \\ 0 & \lambda_2^2 \end{bmatrix}.$$

Extending this product to higher powers, it is readily seen that

$$T^n = CD^nC^{-1}, \tag{4.11}$$

where

$$D^n = \begin{bmatrix} \lambda_1^n & 0 \\ 0 & \lambda_2^n \end{bmatrix} = \begin{bmatrix} 1 & 0 \\ 0 & r^n \end{bmatrix}.$$

The product of the matrices in (4.11) can be expanded to give

$$
\begin{aligned}
T^n = CD^nC^{-1} &= \frac{1}{\alpha + \beta} \begin{bmatrix} 1 & -\alpha \\ 1 & \beta \end{bmatrix} \begin{bmatrix} 1 & 0 \\ 0 & r^n \end{bmatrix} \begin{bmatrix} \beta & \alpha \\ -1 & 1 \end{bmatrix} \\
&= \frac{1}{\alpha + \beta} \begin{bmatrix} 1 & -\alpha r^n \\ 1 & \beta r^n \end{bmatrix} \begin{bmatrix} \beta & \alpha \\ -1 & 1 \end{bmatrix} \\
&= \frac{1}{\alpha + \beta} \begin{bmatrix} \beta & \alpha \\ \beta & \alpha \end{bmatrix} + \frac{r^n}{\alpha + \beta} \begin{bmatrix} \alpha & -\alpha \\ -\beta & \beta \end{bmatrix} \tag{4.12}
\end{aligned}
$$

Since $0 < \alpha, \beta < 1$, it follows that $|r| < 1$, and consequently that $r^n \to 0$ as $n \to \infty$. Hence, from (4.12),

$$T^n \to \frac{1}{\alpha + \beta} \begin{bmatrix} \beta & \alpha \\ \beta & \alpha \end{bmatrix} = Q,$$

say. Further, for *any* initial probability distribution $\mathbf{p}^{(0)}$, $\mathbf{p}^{(n)}$, the distribution over the states after n steps is given by (see (4.6))

$$
\begin{aligned}
\mathbf{p}^{(n)} = \mathbf{p}^{(0)}T^n &= \begin{bmatrix} p_1^{(0)} & p_2^{(0)} \end{bmatrix} T^n \\
&\to \begin{bmatrix} p_1^{(0)} & p_2^{(0)} \end{bmatrix} Q = \frac{1}{\alpha + \beta} \begin{bmatrix} \beta p_1^{(0)} + \beta p_2^{(0)} & \alpha p_1^{(0)} + \alpha p_2^{(0)} \end{bmatrix} \\
&= \begin{bmatrix} \frac{\beta}{\alpha+\beta} & \frac{\alpha}{\alpha+\beta} \end{bmatrix}, \tag{4.13}
\end{aligned}
$$

as $n \to \infty$ since $p_1^{(0)} + p_2^{(0)} = 1$. Note that the limit is independent of $\mathbf{p}^{(0)}$. This limit, usually denoted by $\mathbf{p}$, is known as the *invariant* or *stationary distribution* of the Markov chain, since it is independent of the initial distribution.

If it is known that a stationary distribution does exist then we can take the limits of both sides of

$$\mathbf{p}^{(n)} = \mathbf{p}^{(n-1)}T$$

as $n \to \infty$ to obtain identity

$$\mathbf{p} = \mathbf{p}T.$$

Putting, say, $\mathbf{p} = [a \quad 1-a]$ it is then possible to solve the matrix equation

$$\begin{bmatrix} a & 1-a \end{bmatrix} = \begin{bmatrix} a & 1-a \end{bmatrix} T = \begin{bmatrix} 1-\alpha & \alpha \\ \beta & 1-\beta \end{bmatrix}$$

for a. Thus, comparing elements

$$a = a(1-\alpha) + (1-a)\beta, \qquad 1-a = a\alpha + (1-a)(1-\beta).$$

Both equations give $a = \beta/(\alpha + \beta)$. This method applies to chains with any finite number of states provided that the stationary distribution exists. However, remember the conditions $0 < \alpha, \beta < 1$ imposed at the beginning of this section. If $\alpha = \beta = 1$, then

$$T = \begin{bmatrix} 0 & 1 \\ 1 & 0 \end{bmatrix}$$

and the powers T^n oscillate between

$$\begin{bmatrix} 0 & 1 \\ 1 & 0 \end{bmatrix} \text{ and } \begin{bmatrix} 1 & 0 \\ 0 & 1 \end{bmatrix}$$

so that this Markov chain does not have a stationary distribution. As we have just illustrated, not all chains have invariant distributions. Some have *limiting distributions* that are not invariant but depend on the initial state of the system.

4.4 Powers of the transition matrix for the m-state chain

The method derived for the two-state chain in the previous section can be generalized to m-state chains. We shall sketch the diagonalization method here. Let T be an $m \times m$ stochastic matrix, in other words, a possible transition matrix. The eigenvalues of T are given by the characteristic equation

$$|T - \lambda I_m| = 0, \tag{4.14}$$

in which I_m is the $m \times m$ identity matrix. Assume that the eigenvalues are *distinct*, and denoted by $\lambda_1, \lambda_2, \ldots, \lambda_m$. The corresponding eigenvectors satisfy the equations

$$[T - \lambda_i I_m]\mathbf{r}_i = \mathbf{0}, \quad (i = 1, 2, \ldots, m). \tag{4.15}$$

Construct the matrix

$$C = [\ \mathbf{r}_1 \quad \mathbf{r}_2 \quad \cdots \quad \mathbf{r}_m\ ],$$

which has the eigenvectors as columns. By matrix multiplication

$$
\begin{aligned}
TC = T[\ \mathbf{r}_1 \quad \mathbf{r}_2 \quad \cdots \quad \mathbf{r}_m\ ] &= [\ T\mathbf{r}_1 \quad T\mathbf{r}_2 \quad \cdots \quad T\mathbf{r}_m\ ] \\
&= [\ \lambda_1\mathbf{r}_1 \quad \lambda_2\mathbf{r}_2 \quad \cdots \quad \lambda_m\mathbf{r}_m\ ] \qquad \text{(by (4.15))} \\
&= [\ \mathbf{r}_1 \quad \mathbf{r}_2 \quad \cdots \quad \mathbf{r}_m\ ]
\begin{bmatrix}
\lambda_1 & 0 & \cdots & 0 \\
0 & \lambda_2 & \cdots & 0 \\
\vdots & \vdots & \ddots & \vdots \\
0 & 0 & \cdots & \lambda_m
\end{bmatrix} \\
&= CD
\end{aligned}
\tag{4.16}
$$

where D is the diagonal matrix defined by

$$
D =
\begin{bmatrix}
\lambda_1 & 0 & \cdots & 0 \\
0 & \lambda_2 & \cdots & 0 \\
\vdots & \vdots & \ddots & \vdots \\
0 & 0 & \cdots & \lambda_m
\end{bmatrix}.
\tag{4.17}
$$

Hence, if (4.16) is multiplied on the right by C^{-1}, then

$$T = CDC^{-1}.$$

Powers of T can now be easily found since

$$
\begin{aligned}
T^2 = (CDC^{-1})(CDC^{-1}) &= (CD)(CC^{-1})(DC^{-1}) \\
&= (CD)I_m(DC^{-1}) = CD^2C^{-1}.
\end{aligned}
$$

Similarly, for the general power n,

$$T^n = CD^nC^{-1},
\tag{4.18}$$

where

$$
D^n =
\begin{bmatrix}
\lambda_1^n & 0 & \cdots & 0 \\
0 & \lambda_2^n & \cdots & 0 \\
\vdots & \vdots & \ddots & \vdots \\
0 & 0 & \cdots & \lambda_m^n
\end{bmatrix}.
$$

Although (4.18) is a useful formula for T^n, the algebra involved can be very heavy even for quite modest chains with $m = 4$ or 5, particularly if T contains unknown constants or parameters. On the other hand, mathematical software is now very easy to apply for the numerical calculation of eigenvalues, eigenvectors and powers of matrices for quite large systems.

Example 4.3. Find the eigenvalues and eigenvectors of the stochastic matrix

$$T = \begin{bmatrix} \frac{1}{4} & \frac{1}{2} & \frac{1}{4} \\ \frac{1}{2} & \frac{1}{4} & \frac{1}{4} \\ \frac{1}{4} & \frac{1}{4} & \frac{1}{2} \end{bmatrix}.$$

Construct a formula for T^n, and find $\lim_{n\to\infty} T^n$.

The eigenvalues of T are given by

$$\det(T - \lambda I_3) = \begin{vmatrix} \frac{1}{4} - \lambda & \frac{1}{2} & \frac{1}{4} \\ \frac{1}{2} & \frac{1}{4} - \lambda & \frac{1}{4} \\ \frac{1}{4} & \frac{1}{4} & \frac{1}{2} - \lambda \end{vmatrix}$$

$$= \begin{vmatrix} 1 - \lambda & 1 - \lambda & 1 - \lambda \\ \frac{1}{2} & \frac{1}{4} - \lambda & \frac{1}{4} \\ \frac{1}{4} & \frac{1}{4} & \frac{1}{2} - \lambda \end{vmatrix} \qquad \text{(adding all the rows)}$$

$$= (1 - \lambda) \begin{vmatrix} 1 & 1 & 1 \\ \frac{1}{2} & \frac{1}{4} - \lambda & \frac{1}{4} \\ \frac{1}{4} & \frac{1}{4} & \frac{1}{2} - \lambda \end{vmatrix}$$

$$= (1 - \lambda) \begin{vmatrix} 1 & 1 & 1 \\ \frac{1}{2} & \frac{1}{4} - \lambda & \frac{1}{4} \\ 0 & 0 & \frac{1}{4} - \lambda \end{vmatrix} \qquad \text{(subtracting } \frac{1}{4} \text{ of row 1 from row 3)}$$

$$= (1 - \lambda)(\tfrac{1}{4} - \lambda)(-\tfrac{1}{4} - \lambda) = 0.$$

Define the eigenvalues to be

$$\lambda_1 = 1, \qquad \lambda_2 = \frac{1}{4}, \qquad \lambda_3 = -\frac{1}{4}.$$

Note that a stochastic matrix *always* has a unit eigenvalue.

The eigenvectors $\mathbf{r}_i (i = 1, 2, 3)$ satisfy

$$(T - \lambda_i I_3)\mathbf{r}_i = \mathbf{0}, \qquad (i = 1, 2, 3).$$

Some routine calculations give

$$\mathbf{r}_1 = \begin{bmatrix} 1 \\ 1 \\ 1 \end{bmatrix}, \qquad \mathbf{r}_2 = \begin{bmatrix} -1 \\ -1 \\ 2 \end{bmatrix}, \qquad \mathbf{r}_3 = \begin{bmatrix} 1 \\ -1 \\ 0 \end{bmatrix}.$$

We now let

$$C = \begin{bmatrix} \mathbf{r}_1 & \mathbf{r}_2 & \mathbf{r}_3 \end{bmatrix} = \begin{bmatrix} 1 & -1 & 1 \\ 1 & -1 & -1 \\ 1 & 2 & 0 \end{bmatrix}.$$

Finally, from (4.18)

$$T^n = CD^nC^{-1}$$

$$= \frac{1}{6} \begin{bmatrix} 1 & -1 & 1 \\ 1 & -1 & -1 \\ 1 & 2 & 0 \end{bmatrix} \begin{bmatrix} 1 & 0 & 0 \\ 0 & (\frac{1}{4})^n & 0 \\ 0 & 0 & (-\frac{1}{4})^n \end{bmatrix} \begin{bmatrix} 2 & 2 & 2 \\ -1 & -1 & 2 \\ 3 & -3 & 0 \end{bmatrix}.$$

As $n \to \infty$,

$$T^n \to \frac{1}{6} \begin{bmatrix} 1 & -1 & 1 \\ 1 & -1 & -1 \\ 1 & 2 & 0 \end{bmatrix} \begin{bmatrix} 1 & 0 & 0 \\ 0 & 0 & 0 \\ 0 & 0 & 0 \end{bmatrix} \begin{bmatrix} 2 & 2 & 2 \\ -1 & -1 & 2 \\ 3 & -3 & 0 \end{bmatrix}$$

$$= \begin{bmatrix} \frac{1}{3} & \frac{1}{3} & \frac{1}{3} \\ \frac{1}{3} & \frac{1}{3} & \frac{1}{3} \\ \frac{1}{3} & \frac{1}{3} & \frac{1}{3} \end{bmatrix} = Q,$$

say. □

Suppose now that T is the transition matrix of a 3-state Markov chain, and that the initial probability distribution is $\mathbf{p}^{(0)}$. Then the probability distribution after n steps is

$$\mathbf{p}^{(n)} = \mathbf{p}^{(0)}T^n.$$

The invariant probability distribution $\mathbf{p}$ is

$$\mathbf{p} = \lim_{n \to \infty} \mathbf{p}^{(n)} = \lim_{n \to \infty} \mathbf{p}^{(0)}T^n = \mathbf{p}^{(0)}Q = \begin{bmatrix} \frac{1}{3} & \frac{1}{3} & \frac{1}{3} \end{bmatrix}.$$

The vector $\mathbf{p}$ gives the long-term probability distribution across the three states. In other words, if any snapshot of the system is eventually taken for large n, then the system is equally likely (in this example) to lie in each of the states, and this is independent of the initial distribution $\mathbf{p}^{(0)}$.

Absorbing barriers were referred to in Section 3.1 in the context of random walks. *Absorbing states* are recognizable in Markov chains by a value 1 in a diagonal element of the transition matrix. Since such matrices are stochastic, then all other elements in the same row must be zero. This means that once entered, there is no escape from the absorbing state. For example, in the Markov chain with

$$T = \begin{bmatrix} \frac{1}{2} & \frac{1}{4} & \frac{1}{4} \\ 0 & 1 & 0 \\ \frac{1}{4} & \frac{1}{4} & \frac{1}{2} \end{bmatrix}, \tag{4.19}$$

then the state E_2 is absorbing.

As we illustrated in Section 3.1, diagrams showing transitions between states are particularly helpful. The states are represented by dots with linking directed curves or edges if a transition is possible: if no transition is possible then no directed edge is drawn. Thus, the transition diagram for the three-state chain

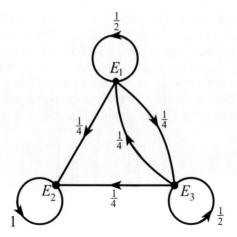

Fig. 4.1 *Transition diagram for the transition matrix in (4.19).*

with a transition matrix given by (4.19) is shown in Figure 4.1. In graph theory terminology, Figure 4.1 shows a *directed graph*. It can be seen that, once entered, there is no escape from the absorbing state E_2.

The eigenvalues of T given by (4.19) are $\lambda_1 = 1$, $\lambda_2 = \frac{1}{4}$, $\lambda_3 = \frac{3}{4}$. The corresponding matrix of eigenvectors is

$$C = \begin{bmatrix} 1 & -1 & 0 \\ 1 & 0 & 0 \\ 1 & 1 & 1 \end{bmatrix}.$$

Using the method illustrated by Example 4.3, it follows that

$$
\begin{aligned}
T^n &= CD^nC^{-1} \\
&= \begin{bmatrix} 1 & -1 & 1 \\ 1 & 0 & 0 \\ 1 & 1 & 1 \end{bmatrix} \begin{bmatrix} 1 & 0 & 0 \\ 0 & (\frac{1}{4})^n & 0 \\ 0 & 0 & (\frac{3}{4})^n \end{bmatrix} \begin{bmatrix} 0 & 1 & 0 \\ -\frac{1}{2} & 0 & \frac{1}{2} \\ \frac{1}{2} & -1 & \frac{1}{2} \end{bmatrix} \\
&\to \begin{bmatrix} 1 & -1 & 1 \\ 1 & 0 & 0 \\ 1 & 1 & 1 \end{bmatrix} \begin{bmatrix} 1 & 0 & 0 \\ 0 & 0 & 0 \\ 0 & 0 & 0 \end{bmatrix} \begin{bmatrix} 0 & 1 & 0 \\ -\frac{1}{2} & 0 & \frac{1}{2} \\ \frac{1}{2} & -1 & \frac{1}{2} \end{bmatrix} \\
&= \begin{bmatrix} 0 & 1 & 0 \\ 0 & 1 & 0 \\ 0 & 1 & 0 \end{bmatrix} = Q,
\end{aligned}
$$

say, as $n \to \infty$. This implies that

$$\mathbf{p} = \lim_{n\to\infty} \mathbf{p}^{(0)} T^n = \mathbf{p}^{(0)} Q = \begin{bmatrix} 0 & 1 & 0 \end{bmatrix}$$

for any initial distribution $\mathbf{p}^{(0)}$: hence it is invariant. This means that the system ultimately ends in E_2 with probability 1.

Example 4.4 (An illness-death model). A possible simple illness-death model can be represented by a four-state Markov chain in which E_1 is a state in which an individual is free of a particular disease, E_2 is a state in which the individual has the disease, and E_3 and E_4 are, respectively, death states arising as a consequence of death as a result of the disease, or from other causes. During some appropriate time interval (perhaps an annual cycle), we assign probabilities to the transition between the states. Suppose that the transition matrix is (in the order of the states),

$$T = \begin{bmatrix} \frac{1}{2} & \frac{1}{4} & 0 & \frac{1}{4} \\ \frac{1}{4} & \frac{1}{2} & \frac{1}{8} & \frac{1}{8} \\ 0 & 0 & 1 & 0 \\ 0 & 0 & 0 & 1 \end{bmatrix}. \tag{4.20}$$

Find the probability that a person ultimately dies from the disease given that he/she did not have the disease initially.

As we might expect this Markov chain has two absorbing states, E_3 and E_4. The individual probabilities can be interpreted as follows, for example: $p_{11} = \frac{1}{2}$ means that an individual, given that he/she is free of the disease in a certain period, has probability $\frac{1}{2}$ of remaining free of the disease; $p_{24} = \frac{1}{8}$ means that the probability that an individual who has the disease but dies from other causes is $\frac{1}{8}$, and so on.

The transition diagram is shown in Figure 4.2.

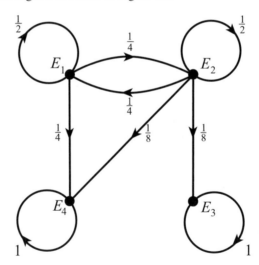

Fig. 4.2 *Transition diagram for the illness–death model.*

In this example it is simpler to *partition* the matrix T as follows. Let

$$T = \begin{bmatrix} A & B \\ O_{22} & I_2 \end{bmatrix},$$

where the *submatrices* A and B are given by

$$A = \begin{bmatrix} \frac{1}{2} & \frac{1}{4} \\ \frac{1}{4} & \frac{1}{2} \end{bmatrix}, \qquad B = \begin{bmatrix} 0 & \frac{1}{4} \\ \frac{1}{8} & \frac{1}{8} \end{bmatrix}, \qquad O_{22} = \begin{bmatrix} 0 & 0 \\ 0 & 0 \end{bmatrix}.$$

Note that A and B are not stochastic matrices. We then observe that

$$T^2 = \begin{bmatrix} A & B \\ O_{22} & I_2 \end{bmatrix} \begin{bmatrix} A & B \\ O_{22} & I_2 \end{bmatrix} = \begin{bmatrix} A^2 & (A + I_2)B \\ O_{22} & I_2 \end{bmatrix},$$

$$T^3 = \begin{bmatrix} A^3 & (A^2 + A + I_2)B \\ O_{22} & I_2 \end{bmatrix},$$

and, in general,

$$T^n = \begin{bmatrix} A^n & (I_2 + A + A^2 + \cdots + A^{n-1})B \\ O_{22} & I_2 \end{bmatrix}.$$

Now let

$$S_n = I_2 + A + \cdots + A^{n-1}.$$

It then follows that

$$(I_2 - A)S_n = (I_2 + A + \cdots + A^{n-1}) - (A + A^2 + \cdots + A^n) = I_2 - A^n,$$

so that

$$S_n = (I_2 - A)^{-1}(I_2 - A^n).$$

Hence

$$T^n = \begin{bmatrix} A^n & (I_2 - A)^{-1}(I_2 - A^n)B \\ O_{22} & I_2 \end{bmatrix}.$$

In the matrix, A is *not* a stochastic matrix, but the method of Section 4.3 to find A^n still works. The eigenvalues of A are given by

$$\det(A - \lambda I_2) = \begin{vmatrix} \frac{1}{2} - \lambda & \frac{1}{4} \\ \frac{1}{4} & \frac{1}{2} - \lambda \end{vmatrix} = (\lambda - \frac{1}{4})(\lambda - \frac{3}{4}) = 0.$$

Let $\lambda_1 = \frac{1}{4}$ and $\lambda_2 = \frac{3}{4}$ (note that there is not a unit eigenvalue in this case since A is not row-stochastic). The matrix C of eigenvectors is

$$C = \begin{bmatrix} -1 & 1 \\ 1 & 1 \end{bmatrix}.$$

Hence

$$T^n = \frac{1}{2} \begin{bmatrix} -1 & 1 \\ 1 & 1 \end{bmatrix} \begin{bmatrix} (\frac{1}{4})^n & 0 \\ 0 & (\frac{3}{4})^n \end{bmatrix} \begin{bmatrix} -1 & 1 \\ 1 & 1 \end{bmatrix}.$$

For the submatrix A, $A^n \to 0$ as $n \to \infty$, so that

$$T^n \to \begin{bmatrix} O_{22} & (I_2 - A)^{-1}B \\ O_{22} & I_2 \end{bmatrix}$$

$$= \begin{bmatrix} 0 & 0 & \frac{1}{6} & \frac{5}{6} \\ 0 & 0 & \frac{1}{3} & \frac{2}{3} \\ 0 & 0 & 1 & 0 \\ 0 & 0 & 0 & 1 \end{bmatrix} = Q,$$

say. In this case

$$\mathbf{p} = \begin{bmatrix} 0 & 0 & \frac{1}{6}p_1^{(0)} + \frac{1}{3}p_2^{(0)} + p_3^{(0)} & \frac{5}{6}p_1^{(0)} + \frac{2}{3}p_2^{(0)} + p_4^{(0)} \end{bmatrix}.$$

This is a *limiting distribution* not an invariant distribution since the result depends on the initial distribution. If $\mathbf{p}^{(0)} = \begin{bmatrix} 1 & 0 & 0 & 0 \end{bmatrix}$, then

$$\mathbf{p} = \begin{bmatrix} 0 & 0 & \frac{1}{6} & \frac{5}{6} \end{bmatrix},$$

from which it can be seen that the probability of an individual not having the disease initially, who subsequently for large n dies as a result of the disease, is $\frac{1}{6}$.
□

The examples we have considered so far have stochastic matrices which have *real* eigenvalues. How do we proceed if the eigenvalues are complex? Consider the following example.

Example 4.5. Find the eigenvalues and eigenvectors of the stochastic matrix

$$T = \begin{bmatrix} \frac{1}{2} & \frac{1}{8} & \frac{3}{8} \\ 1 & 0 & 0 \\ \frac{1}{4} & \frac{1}{2} & \frac{1}{4} \end{bmatrix}.$$

Construct a formula for T^n, and find $\lim_{n\to\infty} T^n$.

The eigenvalues of T are given by

$$\det(T - \lambda I_3) = \begin{vmatrix} \frac{1}{2} - \lambda & \frac{1}{8} & \frac{3}{8} \\ 1 & -\lambda & 0 \\ \frac{1}{4} & \frac{1}{2} & \frac{1}{4} - \lambda \end{vmatrix}$$

$$= (1 - \lambda) \begin{vmatrix} 1 & \frac{1}{8} & \frac{3}{8} \\ 1 & -\lambda & 0 \\ 1 & \frac{1}{2} & \frac{1}{4} - \lambda \end{vmatrix} \quad \text{(adding all the columns)}$$

$$= (1 - \lambda) \begin{vmatrix} 0 & \frac{1}{8} + \lambda & \frac{3}{8} \\ 1 & -\lambda & 0 \\ 0 & \frac{1}{2} + \lambda & \frac{1}{4} - \lambda \end{vmatrix} \quad \begin{array}{l} \text{(subtracting row 2 from rows} \\ \text{1 and 3)} \end{array}$$

$$= (1 - \lambda) \begin{vmatrix} \frac{1}{8} + \lambda & \frac{3}{8} \\ \frac{1}{2} + \lambda & \frac{1}{4} - \lambda \end{vmatrix}$$

$$= (1 - \lambda)(\lambda^2 - \frac{1}{4}\lambda - \frac{3}{16}).$$

Hence the eigenvalues are

$$\lambda_1 = 1, \qquad \lambda_2 = \tfrac{1}{8}(-1 + 3i), \qquad \tfrac{1}{8}(-1 - 3i),$$

two of which are complex conjugates. However, we proceed as before to find the (complex) eigenvectors which are given, as before, by

$$(T - \lambda_i I_3)\mathbf{r}_i = \mathbf{0}, \qquad (i = 1, 2, 3).$$

They can be chosen as

$$\mathbf{r}_1 = \begin{bmatrix} 1 \\ 1 \\ 1 \end{bmatrix}, \qquad \mathbf{r}_2 = \begin{bmatrix} -7 - 9i \\ -16 + 24i \\ 26 \end{bmatrix}, \qquad \mathbf{r}_3 = \begin{bmatrix} -7 + 9i \\ -16 - 24i \\ 26 \end{bmatrix}.$$

We now define C by

$$C = \begin{bmatrix} \mathbf{r}_1 & \mathbf{r}_2 & \mathbf{r}_3 \end{bmatrix} = \begin{bmatrix} 1 & -7 - 9i & -7 + 9i \\ 1 & -16 + 24i & -16 - 24i \\ 1 & 26 & 26 \end{bmatrix}.$$

Inevitably the algebra becomes heavier for the case of complex roots. Calculations can be checked using symbolic computation programs such as *Mathematica* (see the projects for Chapter 4 listed in Chapter 10, and the website for programs in *Mathematica*). The inverse of C is given by

$$C^{-1} = \frac{1}{780} \begin{bmatrix} 416 & 156 & 208 \\ -8 + 14i & -3 - 11i & 11 - 3i \\ -8 - 14i & -3 - 11i & 11 + 3i \end{bmatrix}.$$

The diagonal matrix D of eigenvalues becomes

$$D = \begin{bmatrix} 1 & 0 & 0 \\ 0 & (\frac{-1+3i}{8})^n & 0 \\ 0 & 0 & (\frac{-1-3i}{8})^n \end{bmatrix} \rightarrow \begin{bmatrix} 1 & 0 & 0 \\ 0 & 0 & 0 \\ 0 & 0 & 0 \end{bmatrix}$$

as $n \to \infty$ since $|(-1 \pm 3i)/8| = \sqrt{10}/8 < 1$. Finally, T^n can be calculated from the formula

$$T^n = CD^nC^{-1}.$$

As $n \to \infty$,

$$T^n \rightarrow \begin{bmatrix} \frac{8}{15} & \frac{1}{5} & \frac{4}{15} \\ \frac{8}{15} & \frac{1}{5} & \frac{4}{15} \\ \frac{8}{15} & \frac{1}{5} & \frac{4}{15} \end{bmatrix} = Q.$$

We conclude that complex eigenvalues can be dealt with by the same procedure as real ones: the resulting formulae for T^n and its limit Q will turn out to be real matrices. □

4.5 Gambler's ruin as a Markov chain

We start by summarizing the game (see Section 2.1). It is a game of chance between two players A and B. The gambler A starts with k units (pounds or dollars, etc) and the opponent B with $a - k$ units, where a and k are integers. At each play, A either wins from B one unit with probability p or loses one unit to B with probability $q = 1 - p$. The game ends when either player A or B has no stake. What is the probability that A loses?

The states are $E_0, E_1, \ldots, E_a$ (it is convenient in this application to let the list run from 0 to a), where E_r is the state in which the gambler A has r units. This is a Markov chain, but the transitions are only possible between neighbouring states. This is also the case for the simple random walk. We interpret E_0 and E_a as absorbing states since the game ends when these states are reached. From the rules of the game, the transition matrix is

$$T = \begin{bmatrix} 1 & 0 & 0 & \cdots & 0 & 0 & 0 \\ 1-p & 0 & p & \cdots & 0 & 0 & 0 \\ \vdots & \vdots & \vdots & \ddots & \vdots & \vdots & \vdots \\ 0 & 0 & 0 & \cdots & 1-p & 0 & p \\ 0 & 0 & 0 & \cdots & 0 & 0 & 1 \end{bmatrix} \quad \text{(an } (r+1) \times (r+1) \text{ matrix).}$$

(4.21)

T is an example of a *tridiagonal matrix*.

The diagonalization of T is not really a practical proposition for this $(r+1) \times (r+1)$ matrix except for small r. This is often the case for chains with a large number of states.

The initial distribution $\mathbf{p}^{(0)}$ for the gambler's ruin problem has the elements

$$p_i^{(0)} = \begin{cases} 0 & i \neq k, \\ 1 & i = k, \end{cases} \quad (0 \leq i \leq a, \quad 1 \leq k \leq a - 1),$$

assuming an initial stake of k units. In $\mathbf{p}^{(n)}$, the component $p_0^{(n)}$ is the probability that the gambler loses the game by the nth play, and $p_a^{(n)}$ is the probability that the game is won by the nth play.

Example 4.6. In a gambler's ruin problem suppose that p is the probability that the gambler loses at each play, and that $a = 5$ and that the gambler's initial stake is $k = 3$ units. Compute the probability that the gambler loses/wins by the fourth play. What is the probability that the gambler actually loses at the fifth play.

In this example, the transition matrix (4.21) is the 6×6 matrix

$$T = \begin{bmatrix} 1 & 0 & 0 & 0 & 0 & 0 \\ 1-p & 0 & p & 0 & 0 & 0 \\ 0 & 1-p & 0 & p & 0 & 0 \\ 0 & 0 & 1-p & 0 & p & 0 \\ 0 & 0 & 0 & 1-p & 0 & p \\ 0 & 0 & 0 & 0 & 0 & 1 \end{bmatrix},$$

and

$$\mathbf{p}^{(0)} = [\; 0 \quad 0 \quad 0 \quad 1 \quad 0 \quad 0 \;].$$

Then

$$\mathbf{p}^{(4)} = \mathbf{p}^{(0)} T^4$$
$$= [\; (1-p)^3 \quad 3(1-p)^3 \quad 0 \quad 5(1-p)^2 p^2 \quad 0 \quad p^2 + 2(1-p)p^3 \;]$$

obtained from sequences of independent events, or by using symbolic computation. From this distribution, the probability that the gambler loses by the fourth play is

$$p_0^{(4)} = (1-p)^3,$$

which must occur as the probability of three successive losses. The probability that the gambler wins is

$$p_5^{(4)} = p^2 + 2(1-p)p^3,$$

derived from two successive wins, or from the sequences

win/lose/win/win or *lose/win/win/win*

which together give the term $2(1-p)p^3$. The probability that the gambler actually wins *at* the fourth play is $2(1-p)p^3$. $\qquad\square$

For the general $(a+1) \times (a+1)$ transition matrix (4.21) for the gambler's ruin, whilst T^n is difficult to obtain, $Q = \lim_{n\to} T^n$ can be found if we assume that the shape of the limiting matrix is

$$Q = \begin{bmatrix} 1 & 0 & \cdots & 0 & 0 \\ u_1 & 0 & \cdots & 0 & 1-u_1 \\ u_2 & 0 & \cdots & 0 & 1-u_2 \\ \vdots & \vdots & \ddots & \vdots & \vdots \\ u_{a-1} & 0 & \cdots & 0 & 1-u_{a-1} \\ 0 & 0 & \cdots & 0 & 1 \end{bmatrix}.$$

The implication of the zeros in columns 2 to $a-1$ is that, eventually, the chain must end in either state E_0 or E_a (see Section 2.2). We then observe that, after rearranging indices,

$$Q = \lim_{n\to\infty} T^{n+1} = (\lim_{n\to\infty} T^n)T = QT, \qquad (4.22)$$

and

$$Q = \lim_{n\to\infty} T^{n+1} = T(\lim_{n\to\infty} T^n) = TQ. \qquad (4.23)$$

In this case, (4.22) turns out to be an identity. However (4.23) implies

$$
TQ - Q = (T - I_{a-1})Q =
$$

$$
\begin{bmatrix}
0 & 0 & 0 & \cdots & 0 & 0 & 0 \\
1-p & -1 & p & \cdots & 0 & 0 & 0 \\
0 & 1-p & -1 & \cdots & 0 & 0 & 0 \\
\vdots & \vdots & \vdots & \ddots & \vdots & \vdots & \vdots \\
0 & 0 & 0 & \cdots & 1-p & -1 & p \\
0 & 0 & 0 & \cdots & 0 & 0 & 0
\end{bmatrix}
\begin{bmatrix}
1 & 0 & \cdots & 0 & 0 \\
u_1 & 0 & \cdots & 0 & 1-u_1 \\
u_2 & 0 & \cdots & 0 & 1-u_2 \\
\vdots & \vdots & \ddots & \vdots & \vdots \\
u_{a-1} & 0 & \cdots & 0 & 1-u_{a-1} \\
0 & 0 & \cdots & 0 & 1
\end{bmatrix}
$$

$$
=
\begin{bmatrix}
0 & 0 & \cdots & 0 & 0 \\
(1-p) - u_1 + pu_2 & 0 & \cdots & 0 & -(1-p) + u_1 - pu_2 \\
(1-p)u_1 - u_2 + pu_3 & 0 & \cdots & 0 & -(1-p)u_1 + u_2 - pu_3 \\
\vdots & \vdots & \ddots & \vdots & \vdots \\
(1-p)u_{a-2} - u_{a-1} & 0 & \cdots & 0 & -(1-p)u_{a-2} + u_{a-1} \\
0 & 0 & \cdots & 0 & 0
\end{bmatrix}
$$

This equation is zero, implying $TQ = Q$ if the elements in the first and last columns are all zero, but note that the elements in the last column are simply the ones in the first column multiplied by -1. The result is

$$
pu_2 - u_1 + (1-p) = 0,
$$

$$
pu_{k+2} - u_{k+1} + (1-p)u_k = 0, \qquad (k = 1, 2, \ldots, a-2),
$$

$$
-u_{a-1} + (1-p)u_{a-2} = 0.
$$

This is equivalent to writing

$$
pu_{k+2} - u_{k+1} + (1-p)u_k = 0, \qquad (k = 0, 1, 2, \ldots, a-2),
$$

$$
u_0 = 1, \qquad u_a = 0,
$$

the latter two equations now defining the boundary conditions. These are the difference equations and boundary conditions for the gambler's ruin problem derived using the law of total probability in equations (2.2) and (2.3).

4.6 Classification of states

Let us return to the general m-state chain with states $E_1, E_2, \ldots, E_m$ and transition matrix

$$
T = [p_{ij}], \qquad (1 \le i, j \le m).
$$

For a homogeneous chain, recollect that p_{ij} is the probability that a transition occurs between E_i and E_j at any step or change of state in the chain. We intend to investigate and classify some of the more common types of states that can occur in Markov chains. This will be a brief treatment, using mainly examples of what is an extensive algebraic problem.

(a) Absorbing state

We have already met one state – namely the *absorbing state*. Once entered there is no escape from an absorbing state. An absorbing state E_i is characterized by

$$p_{ii} = 1, \qquad p_{ij} = 0, (i \neq j, \quad j = 1, 2, \ldots m),$$

in the ith row of T.

(b) Periodic state

The probability of a return to E_i at step n is $p_{ii}^{(n)}$. Let t be an integer greater than 1. Suppose that

$$p_{ii}^{(n)} = 0 \text{ for } n \neq t, 2t, 3t, \ldots$$

$$p_{ii}^{(n)} \neq 0 \text{ for } n = t, 2t, 3t, \ldots.$$

In this case the state E_i is said to be *periodic* with period t. If, for a state, no such t exists with this property, then the state is described as *aperiodic*. Let

$$d(i) = \gcd\{n \,|\, p_{ii}^{(n)} > 0\},$$

that is, the greatest common divisor (gcd) of the set of integers n for which for which $p_{ii}^{(n)} > 0$. Then the state E_i is said to be *periodic* if $d(i) > 1$ and *aperiodic* if $d(i) = 1$. The definition above is included in this. See Problem 4.12 for an example of such periodic states that satisfy this general definition but not the one above.)

Example 4.7. A four-state Markov chain has the transition matrix

$$T = \begin{bmatrix} 0 & \frac{1}{2} & 0 & \frac{1}{2} \\ 0 & 0 & 1 & 0 \\ 1 & 0 & 0 & 0 \\ 0 & 0 & 1 & 0 \end{bmatrix}.$$

Show that all states have period 3.

The transition diagram is shown in Figure 4.3, from which it is clear that all states are period 3. For example, if the chain starts in E_1, then returns to E_1 are only possible at steps 3, 6, 9, ... either through E_2 or E_3.

The analysis of chains with periodic states can be complicated. However, one can check for a suspected periodicity as follows. By direct computation

$$S = T^3 = \begin{bmatrix} 1 & 0 & 0 & 0 \\ 0 & \frac{1}{2} & 0 & \frac{1}{2} \\ 0 & 0 & 1 & 0 \\ 0 & \frac{1}{2} & 0 & \frac{1}{2} \end{bmatrix}.$$

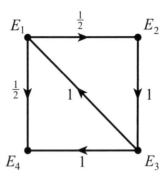

Fig. 4.3 *Transition diagram for Example 4.7.*

In this example,

$$S^2 = T^6 = SS = S,$$

so that

$$S^r = T^{3r} = S, \qquad (r = 1, 2, \ldots),$$

which always has non-zero elements on its diagonal. On the other hand

$$S^{r+1} = S^r S = \begin{bmatrix} 0 & \frac{1}{2} & 0 & \frac{1}{2} \\ 0 & 0 & 1 & 0 \\ 1 & 0 & 0 & 0 \\ 0 & 0 & 1 & 0 \end{bmatrix}, \qquad S^{r+2} = S^r S^2 = \begin{bmatrix} 0 & 0 & 1 & 0 \\ 1 & 0 & 0 & 0 \\ 0 & \frac{1}{2} & 0 & \frac{1}{2} \\ 1 & 0 & 0 & 0 \end{bmatrix},$$

and both these matrices have zero diagonal elements for $r = 1, 2, 3, \ldots$. Hence, for $i = 1, 2, 3, 4$,

$$p_{ii}^{(n)} = 0 \text{ for } n \neq 3, 6, 9, \ldots,$$

$$p_{ii}^{(n)} \neq 0 \text{ for } n = 3, 6, 9, \ldots,$$

which means that all states are period 3. □

(c) Persistent state

Let $f_j^{(n)}$ be the probability that the *first return* or *visit* to E_j occurs at the nth step. This probability is not the same as $p_{jj}^{(n)}$, which is the probability that a return occurs at the nth step, and includes possible returns at steps $1, 2, 3, \ldots, n-1$ also. It follows that

$$p_{jj}^{(1)} (= p_{jj}) = f_j^{(1)}, \tag{4.24}$$

$$p_{jj}^{(2)} = f_j^{(2)} + f_j^{(1)} p_{jj}^{(1)}, \tag{4.25}$$

$$p_{jj}^{(3)} = f_j^{(3)} + f_j^{(1)} p_{jj}^{(2)} + f_j^{(2)} p_{jj}^{(1)}, \tag{4.26}$$

and, in general,

$$p_{jj}^{(n)} = f_j^{(n)} + \sum_{r=1}^{n-1} f_j^{(r)} p_{jj}^{(n-r)} \qquad (n \geq 2). \tag{4.27}$$

In words, what (4.26) means is that the probability of a return at the third step is the probability of a first return at the third step, or the probability of a first return at the first step and a return two steps later, or the probability of a first return at the second step and a return one step later.

Equations (4.24) and (4.27) become iterative formulas for the sequence of first returns $f_j^{(n)}$ which can be expressed as:

$$f_j^{(1)} = p_{jj}, \tag{4.28}$$

$$f_j^{(n)} = p_{jj}^{(n)} - \sum_{r=1}^{n-1} f_j^{(r)} p_{jj}^{(n-r)} \qquad (n \geq 2). \tag{4.29}$$

The probability that a chain returns at some step to the state E_j is

$$f_j = \sum_{n=1}^{\infty} f_j^{(n)}.$$

If $f_j = 1$, then a return to E_j is certain, and E_j is called a *persistent state*.

Example 4.8. A three-state Markov chain has the transition matrix

$$T = \begin{bmatrix} p & 1-p & 0 \\ 0 & 0 & 1 \\ 1-q & 0 & q \end{bmatrix},$$

where $0 < p < 1, 0 < q < 1$. Show that the state E_1 is persistent.

For simple chains, a direct approach using the transition diagram is often easier than the formula (4.29) for $f_j^{(n)}$. For this example, the transition diagram is shown in Figure 4.4. If a sequence starts in E_1, then it can be seen that *first returns* to E_1 can be made to E_1 at every step except for $n = 2$, since after two steps the chain must be in state E_3. From the figure it can be argued that

$$f_1^{(1)} = p, \qquad f_1^{(2)} = 0, \qquad f_1^{(3)} = (1-p).1.(1-q),$$

$$f_1^{(n)} = (1-p).1.q^{n-3}.(1-q), \qquad (n \geq 3).$$

For $n \geq 3$, the sequence giving $f_1^{(n)}$ will be, for $n \geq 4$,

$$E_1 \; E_2 \; \overbrace{E_3 \; E_3 \; \cdots \; E_3}^{(n-3) \text{ times}} \; E_1.$$

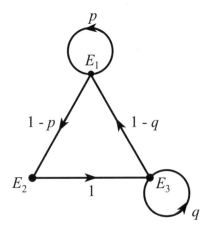

Fig. 4.4 *Transition diagram for Example 4.8.*

The probability f_1 that the system returns at least once to E_1 is

$$f_1 = \sum_{n=1}^{\infty} f_1^{(n)} = p + \sum_{n=3}^{\infty} (1-p)(1-q)q^{n-3},$$

$$= p + (1-p)(1-q) \sum_{s=0}^{\infty} q^s, \qquad (s = n-3)$$

$$= p + (1-p)\frac{(1-q)}{(1-q)} = 1,$$

using the sum formula for the geometric series. Hence $f_1 = 1$, and consequently the state E_1 is persistent.

In this example, it is almost self-evident that all states in this chain will be persistent since there is no escape route to an absorbing state, for example. □

Persistent states can have a further distinction. The *mean recurrence time* μ_j of a persistent state E_j, for which $\sum_{n=1}^{\infty} f_j^{(n)} = 1$, is given by

$$\mu_j = \sum_{n=1}^{\infty} n f_j^{(n)}. \tag{4.30}$$

In Example 4.8 above, the state E_1 is persistent and its mean recurrence time is given by

$$\mu_1 = \sum_{n=1}^{\infty} n f_1^{(n)} = p + (1-p)(1-q) \sum_{n=3}^{\infty} n q^{n-3}$$

$$= \frac{3 - 2p - 2q + pq}{1-q},$$

which is finite. For some chains, however, the mean recurrence time can be infinite; in other words, the mean number of steps to a first return is unbounded.

A persistent state E_j is said to be *null* if $\mu_j = \infty$, and *non-null* if $\mu_j < \infty$. All states in Example 4.8 are non-null persistent.

To create a simple example of a finite chain with a persistent null state, we consider a chain in which the transition probabilities depend on the step number n. Null states are not possible in finite chains with a *constant* transition matrix.

Example 4.9. A three-state inhomogeneous Markov chain has the transition matrix

$$
T_n = \begin{bmatrix} \frac{1}{2} & \frac{1}{2} & 0 \\ 0 & 0 & 1 \\ 1/(n+1) & 0 & n/(n+1) \end{bmatrix},
$$

where T_n is the transition matrix at step n. Show that E_1 is a persistent null state.

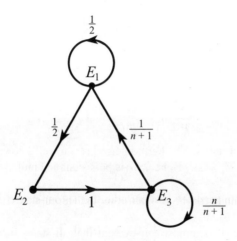

Fig. 4.5 *Transition diagram for Example 4.9.*

The transition diagram at a general step n is shown in Figure 4.5. From the figure

$$
f_1^{(1)} = \frac{1}{2}, \qquad f_1^{(2)} = 0, \qquad f_1^{(3)} = \frac{1}{2} \cdot 1 \cdot \frac{1}{4},
$$

$$
f_1^{(n)} = \frac{1}{2} \cdot 1 \cdot \frac{3}{4} \cdot \frac{4}{5} \cdots \frac{n-1}{n} \cdot \frac{1}{n+1} = \frac{3}{2n(n+1)}, \qquad (n \geq 4).
$$

Hence

$$
f_1 = \frac{1}{2} + \frac{1}{8} + \frac{3}{2} \sum_{n=4}^{\infty} \frac{1}{n(n+1)}.
$$

Since

$$
\frac{1}{n(n+1)} = \frac{1}{n} - \frac{1}{n+1},
$$

it follows that

$$\sum_{n=4}^{\infty} \frac{1}{n(n+1)} = \lim_{N \to \infty} \sum_{n=4}^{N} \left(\frac{1}{n} - \frac{1}{n+1} \right) = \lim_{N \to \infty} \left(\frac{1}{4} - \frac{1}{N+1} \right) = \frac{1}{4}.$$

Hence

$$f_1 = \frac{5}{8} + \frac{3}{8} = 1,$$

which means that E_1 is persistent. On the other hand, the mean recurrence time

$$\mu_1 = \sum_{n=1}^{\infty} n f_1^{(n)} = \frac{7}{8} + \frac{3}{2} \sum_{n=4}^{\infty} \frac{n}{n(n+1)},$$

$$= \frac{7}{8} + \frac{3}{2} \left(\frac{1}{5} + \frac{1}{6} + \frac{1}{7} + \cdots \right),$$

$$= \frac{7}{8} + \frac{3}{2} \sum_{n=5}^{\infty} \frac{1}{n}.$$

The series in the previous equation is the *harmonic series*

$$\sum_{n=1}^{\infty} \frac{1}{n} = 1 + \frac{1}{2} + \frac{1}{3} + \cdots,$$

minus the first four terms. The harmonic series is a well-known divergent series, which means that $\mu_1 = \infty$. Hence, E_1 is persistent and null. $\quad\square$

States can be both persistent and periodic. In the four-state chain with transition matrix

$$T = \begin{bmatrix} 0 & 1 & 0 & 0 \\ 0 & 0 & 0 & 1 \\ 0 & 1 & 0 & 0 \\ \frac{1}{2} & 0 & \frac{1}{2} & 0 \end{bmatrix},$$

all states are period 3, persistent and non-null.

(d) Transient state

For a persistent state the probability of a first return at some step in the future is certain. For some states

$$f_j = \sum_{n=1}^{\infty} f_j^{(n)} < 1, \tag{4.31}$$

which means that the probability of a first return is *not* certain. Such states are described as *transient*.

Example 4.10. A four-state Markov chain has the transition matrix

$$T = \begin{bmatrix} 0 & \frac{1}{2} & \frac{1}{4} & \frac{1}{4} \\ \frac{1}{2} & \frac{1}{2} & 0 & 0 \\ 0 & 0 & 1 & 0 \\ 0 & 0 & \frac{1}{2} & \frac{1}{2}. \end{bmatrix}$$

Show that E_1 is a transient state.

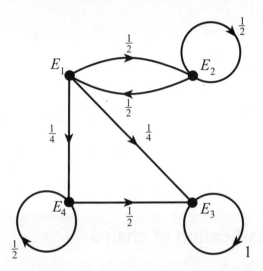

Fig. 4.6 *Transition diagram for Example 4.10.*

The transition diagram is shown in Figure 4.6. From the figure

$$f_1^{(1)} = 0, \quad f_1^{(2)} = \frac{1}{2} \cdot \frac{1}{2} = (\tfrac{1}{2})^2, \quad f_1^{(3)} = (\tfrac{1}{2})^3, \quad f_1^{(n)} = (\tfrac{1}{2})^n.$$

Hence

$$f_1 = \sum_{n=1}^{\infty} f_1^{(n)} = \sum_{n=2}^{\infty} (\tfrac{1}{2})^n = \tfrac{1}{2} < 1.$$

Hence E_1 is a transient state. The reason for the transience of E_1 can be seen from Figure 4.6, where transitions from E_3 or E_4 to E_1 or E_2 are not possible. □

(e) Ergodic state

An important state, which we will consider in the next section, is the state that is persistent, non-null and aperiodic. This state is called *ergodic*. Ergodic states are important in the classification of chains, and in proving the existence of limiting probability distributions, as we shall see in the following section.

Example 4.11. In Example 4.8 we considered the three-state Markov chain with transition matrix

$$T = \begin{bmatrix} p & 1-p & 0 \\ 0 & 0 & 1 \\ 1-q & 0 & q \end{bmatrix}$$

where $0 < p < 1, 0 < q < 1$. Show that state E_1 is ergodic.

It was shown in Example 4.8 that E_1 was persistent with

$$f_1^{(1)} = p, \qquad f_1^{(2)} = 0, \qquad f_1^{(n)} = (1-p)(1-q)q^{n-3} \qquad (n \geq 3).$$

It follows that its mean recurrence time is

$$\mu_1 = \sum_{n=1}^{\infty} n f_1^{(n)} = p + (1-p)(1-q) \sum_{n=3}^{\infty} n q^{n-3} = \frac{3 - 2q}{(1-q)^2} < \infty.$$

The convergence of μ_1 implies that E_1 is non-null. In addition, the diagonal elements $p_{ii}^{(n)} > 0$ for $n \geq 3$ and $i = 1, 2, 3$, which means that E_1 is aperiodic. Hence, from the definition above E_1 (and also E_2 and E_3) is ergodic. $\qquad\square$

4.7 Classification of chains

In the previous section, we considered some defining properties of individual states. In this section we define properties of chains that are common properties of states in the chain.

(a) Irreducible chains

An *irreducible chain* is one in which every state can be *reached* or is *accessible* from every other state in the chain in a finite number of steps. That any state E_j can be reached from any other state E_i means that $p_{ij}^{(n)} > 0$ for some integer n.

A matrix $A = [a_{ij}]$ is said to be *positive* if $a_{ij} > 0$ for all i, j. A Markov chain with transition matrix T is said to be *regular* if there exists an integer N such that T^N is positive.

A regular chain is obviously irreducible. However, the converse is not necessarily true, as can be seen from the simple two-state chain with transition matrix

$$T = \begin{bmatrix} 0 & 1 \\ 1 & 0 \end{bmatrix}.$$

Since

$$T^{2n} = \begin{bmatrix} 1 & 0 \\ 0 & 1 \end{bmatrix} = I_2 \quad \text{and} \quad T^{2n+1} = \begin{bmatrix} 0 & 1 \\ 1 & 0 \end{bmatrix} = T,$$

for $n = 1, 2, 3 \ldots.$. No power of T is a positive matrix.

Example 4.12. Show that the three-state chain with transition matrix

$$T = \begin{bmatrix} \frac{1}{3} & \frac{1}{3} & \frac{1}{3} \\ 0 & 0 & 1 \\ 1 & 0 & 0 \end{bmatrix}$$

defines a regular (and hence irreducible) chain.

For the transition matrix T

$$T^2 = \begin{bmatrix} \frac{4}{9} & \frac{1}{9} & \frac{4}{9} \\ 1 & 0 & 0 \\ \frac{1}{3} & \frac{1}{3} & \frac{1}{3} \end{bmatrix}, \qquad T^3 = \begin{bmatrix} \frac{16}{27} & \frac{4}{27} & \frac{7}{27} \\ \frac{1}{3} & \frac{1}{3} & \frac{1}{3} \\ \frac{4}{9} & \frac{1}{9} & \frac{4}{9} \end{bmatrix}.$$

Hence T^3 is a positive matrix, which means that the chain is regular. □

An important feature of an irreducible chain is that all its states are of the same type, that is, either all transient or all persistent (null or non-null), and all have the same period. A proof of this is given by Feller (1968, p. 391). This means that the classification of all states in an irreducible chain can be inferred from the known classification of one state. It is intuitively reasonable to infer also that the states of a finite irreducible chain cannot all be transient, since it would mean that a return to *any* state would not be certain, even though all states are accessible from all other states in a finite number of steps. This requires a proof that will not be included here.

(b) Closed sets

A Markov chain may contain some states that are transient, some that are persistent, absorbing states, and so on. The persistent states can be part of closed subchains. A set of states C in a Markov chain is said to *closed* if any state within C can be reached from any other state within C, and no state outside C can be reached from any state inside C. Algebraically, a *necessary* condition for this to be the case is that

$$p_{ij} = 0 \quad \forall E_i \in C \quad \text{and} \quad \forall E_j \notin C.$$

An absorbing state is closed with just one state. Note also that a closed subset is itself an irreducible subchain of the full Markov chain.

Example 4.13. Discuss the status of each state in the six-state Markov chain with transition matrix

$$
T = \begin{bmatrix}
\frac{1}{2} & \frac{1}{2} & 0 & 0 & 0 & 0 \\
\frac{1}{4} & \frac{3}{4} & 0 & 0 & 0 & 0 \\
\frac{1}{4} & \frac{1}{4} & \frac{1}{4} & \frac{1}{4} & 0 & 0 \\
\frac{1}{4} & 0 & \frac{1}{4} & \frac{1}{4} & 0 & \frac{1}{4} \\
0 & 0 & 0 & 0 & \frac{1}{2} & \frac{1}{2} \\
0 & 0 & 0 & 0 & \frac{1}{2} & \frac{1}{2}
\end{bmatrix}.
$$

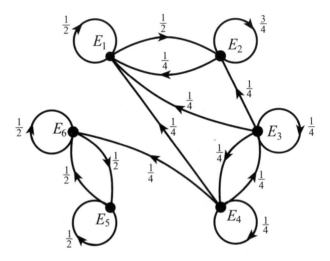

Fig. 4.7 *Transition diagram for Example 4.13.*

A diagram representing the chain is shown in Figure 4.7 . As usual, the figure is a great help in settling questions of which sets of states are closed. It can be seen that $\{E_1, E_2\}$ is a closed irreducible subchain since no states outside the states can be reached from E_1 and E_2. Similarly, $\{E_5, E_6\}$ is a closed irreducible subchain. The states E_3 and E_4 are transient. All states are aperiodic, which means that E_1, E_2, E_5 and E_6 are ergodic.

(c) Ergodic chains

As we have seen, all the states in an irreducible chain belong to the same class. If all states are ergodic, that is, persistent, non-null and aperiodic, then the chain is described as an *ergodic chain*.

Example 4.14. Show that all states of the chain with transition matrix

$$T = \begin{bmatrix} \frac{1}{3} & \frac{1}{3} & \frac{1}{3} \\ 0 & 0 & 1 \\ 1 & 0 & 0 \end{bmatrix}$$

are ergodic.

This is the same chain as in Example 4.12 where it was shown to be irreducible and regular, which means that all states must be persistent, non-null and aperiodic. Hence all states are ergodic. □

Example 4.15. Consider the three-state Markov chain with transition matrix

$$T = \begin{bmatrix} \frac{1}{5} & \frac{4}{5} & 0 \\ 0 & 0 & 1 \\ 1 & 0 & 0 \end{bmatrix}.$$

Show that all states are ergodic. Find the eigenvalues of T, and $Q = \lim_{n \to \infty} T^n$. Determine the mean recurrence times μ_1, μ_2, μ_3 for each state, and confirm that the rows of Q all have the elements $1/\mu_1, 1/\mu_2, 1/\mu_3$.

It is easy to check that T^4 is a positive matrix, which implies that the chain is ergodic. The eigenvalues of T are given by

$$\det(T - \lambda I_3) = \begin{vmatrix} \frac{1}{5} - \lambda & \frac{4}{5} & 0 \\ 0 & -\lambda & 1 \\ 1 & 0 & -\lambda \end{vmatrix}$$

$$= -\lambda^3 + \frac{1}{5}\lambda^2 + \frac{4}{5}$$

$$= \frac{1}{5}(1 - \lambda)(5\lambda^2 + 4\lambda + 4) = 0.$$

Hence the eigenvalues can be denoted by

$$\lambda_1 = 1, \qquad \lambda_2 = -\tfrac{2}{5} + \tfrac{4}{5}i, \qquad \lambda_3 = -\tfrac{2}{5} - \tfrac{4}{5}i.$$

The corresponding eigenvectors are

$$\mathbf{r}_1 = \begin{bmatrix} 1 \\ 1 \\ 1 \end{bmatrix}, \qquad \mathbf{r}_2 = \begin{bmatrix} -\frac{2}{5} + \frac{4}{5}i \\ -\frac{1}{2} - i \\ 1 \end{bmatrix}, \qquad \mathbf{r}_3 = \begin{bmatrix} -\frac{2}{5} - \frac{4}{5}i \\ -\frac{1}{2} + i \\ 1 \end{bmatrix}.$$

Let

$$C = \begin{bmatrix} \mathbf{r}_1 & \mathbf{r}_2 & \mathbf{r}_3 \end{bmatrix} = \begin{bmatrix} 1 & -\frac{2}{5} + \frac{4}{5}i & -\frac{2}{5} - \frac{4}{5}i \\ 1 & -\frac{1}{2} - i & -\frac{1}{2} + i \\ 1 & 1 & 1 \end{bmatrix}.$$

The computed inverse is given by

$$C^{-1} = \frac{1}{52} \begin{bmatrix} 20 & 16 & 16 \\ -10-30i & -8+14i & 18+i \\ -10+30i & -8-14i & 18-i \end{bmatrix}.$$

As in Example 4.5, it follows that

$$Q = \lim_{n \to \infty} T^n = C \begin{bmatrix} 1 & 0 & 0 \\ 0 & 0 & 0 \\ 0 & 0 & 0 \end{bmatrix} C^{-1} = \frac{1}{13} \begin{bmatrix} 5 & 4 & 4 \\ 5 & 4 & 4 \\ 5 & 4 & 4 \end{bmatrix}.$$

The invariant dustribution is therefore $\mathbf{p} = \begin{bmatrix} \frac{5}{13} & \frac{4}{13} & \frac{4}{13} \end{bmatrix}$. Note that the elements in $\mathbf{p}$ are the same as the *first row* in C^{-1}. Is this always the case for ergodic chains if the first eigenvalue $\lambda_1 = 1$?

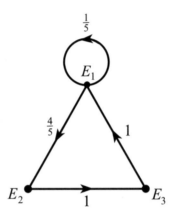

Fig. 4.8 *Transition diagram for Example 4.15.*

The first returns $f_i^{(n)}$ for each of the states can be easily calculated from the transition diagram Figure 4.8. Thus

$$f_1^{(1)} = \frac{1}{5}, \qquad f_1^{(2)} = 0, \qquad f_1^{(3)} = \frac{4}{5}.1.1 = \frac{4}{5},$$

$$f_2^{(1)} = f_3^{(1)} = 0, \quad f_2^{(2)} = f_3^{(2)} = 0, \quad f_2^{(n)} = f_2^{(n)} = \frac{4}{5}\left(\frac{1}{5}\right)^{n-3}, \quad (n \geq 3).$$

Hence, by equation (4.30),

$$\mu_1 = \sum_{n=1}^{\infty} n f_1^{(n)} = \frac{1}{5} + 3.\frac{4}{5} = \frac{13}{5}.$$

$$\mu_2 = \mu_3 = \sum_{n=1}^{\infty} n f_2^{(n)} = \frac{4}{5} \sum_{n=3}^{\infty} n(\frac{1}{5})^{n-3} = \frac{13}{4}.$$

The vector of reciprocals

$$\begin{bmatrix} \frac{1}{\mu_1} & \frac{1}{\mu_2} & \frac{1}{\mu_3} \end{bmatrix} = \begin{bmatrix} \frac{5}{13} & \frac{4}{13} & \frac{4}{13} \end{bmatrix}$$

agrees with the vector $\mathbf{p}$ above calculated by the eigenvalue method. $\qquad\square$

For ergodic chains this is always the case: the *invariant distribution is the vector of mean recurrence time reciprocals.*

Problems

4.1. If $T = [p_{ij}]$, $(i, j = 1, 2, 3)$ and

$$p_{ij} = \frac{i + j}{6 + 3i},$$

show that T is a row-stochastic matrix. What is the probability that a transition between states E_2 and E_3 occurs at any step?

If the initial probability distribution in a Markov chain is

$$\mathbf{p}^{(0)} = \begin{bmatrix} \frac{1}{2} & \frac{1}{4} & \frac{1}{4} \end{bmatrix},$$

what are the probabilities that states E_1, E_2 and E_3 are occupied after one step. Explain why the probability that the chain finishes in state E_2 is $\frac{1}{3}$ irrespective of the number of steps.

4.2. If

$$T = \begin{bmatrix} \frac{1}{2} & \frac{1}{4} & \frac{1}{4} \\ \frac{1}{3} & \frac{1}{3} & \frac{1}{3} \\ \frac{1}{4} & \frac{1}{2} & \frac{1}{4} \end{bmatrix},$$

calculate $p_{22}^{(2)}$, $p_{31}^{(2)}$ and $p_{13}^{(2)}$.

4.3. For the transition matrix

$$T = \begin{bmatrix} \frac{1}{3} & \frac{2}{3} \\ \frac{1}{4} & \frac{3}{4} \end{bmatrix}$$

calculate $p_{12}^{(3)}$, $p_2^{(3)}$ and $\mathbf{p}^{(3)}$ given that $\mathbf{p}^{(0)} = \begin{bmatrix} \frac{1}{2} & \frac{1}{2} \end{bmatrix}$. Also, find the eigenvalues of T, construct a formula for T^n and obtain $\lim_{n \to \infty} T^n$.

4.4. Sketch transition diagrams for each of the following three-state Markov chains.

(a) $A = \begin{bmatrix} \frac{1}{3} & \frac{1}{3} & \frac{1}{3} \\ 0 & 0 & 1 \\ 1 & 0 & 0 \end{bmatrix}$; (b) $B = \begin{bmatrix} \frac{1}{2} & \frac{1}{4} & \frac{1}{4} \\ 0 & 1 & 0 \\ \frac{1}{2} & \frac{1}{2} & 0 \end{bmatrix}$; (c) $C = \begin{bmatrix} 0 & \frac{1}{2} & \frac{1}{2} \\ 1 & 0 & 0 \\ \frac{1}{3} & \frac{1}{3} & \frac{1}{3} \end{bmatrix}$.

4.5. Find the eigenvalues of

$$T = \begin{bmatrix} a & b & c \\ c & a & b \\ b & c & a \end{bmatrix}, \quad (a > 0, b > 0, c > 0).$$

Show that the eigenvalues are complex if $b \neq c$. (If $a + b + c = 1$, then T is a *doubly-stochastic matrix*.)

Find the eigenvalues and eigenvectors in the following cases:

(a) $a = \frac{1}{2}, b = \frac{1}{4}, c = \frac{1}{4}$;

(b) $a = \frac{1}{2}, b = \frac{1}{8}, c = \frac{3}{8}$.

4.6. Find the eigenvalues, eigenvectors, the matrix of eigenvectors C, its inverse C^{-1}, a formula for T^n and $\lim_{n \to \infty} T^n$ for each of the following transition matrices;

(a)

$$T = \begin{bmatrix} \frac{1}{8} & \frac{7}{8} \\ \frac{1}{2} & \frac{1}{2} \end{bmatrix};$$

(b)

$$T = \begin{bmatrix} \frac{1}{2} & \frac{1}{8} & \frac{3}{8} \\ \frac{1}{4} & \frac{3}{8} & \frac{3}{8} \\ \frac{1}{4} & \frac{5}{8} & \frac{1}{8} \end{bmatrix};$$

(c)

$$T = \begin{bmatrix} \frac{1}{4} & \frac{1}{8} & \frac{3}{8} & \frac{1}{4} \\ \frac{1}{3} & \frac{1}{6} & \frac{1}{6} & \frac{1}{3} \\ \frac{1}{3} & \frac{1}{3} & 0 & \frac{1}{3} \\ \frac{1}{3} & 0 & 0 & \frac{2}{3} \end{bmatrix}.$$

4.7. The weather in a certain region can be characterized as being sunny (S), cloudy (C) or rainy (R) on any particular day. The probability of any type of weather on one day depends only on the state of the weather on the previous day. For example, if it is sunny one day then sun or clouds are equally likely on the next day with no possibility of rain. Explain what other the day-to-day possibilities are if the weather is represented by the transition matrix

$$T = \begin{array}{c|ccc} & S & C & R \\ \hline S & \frac{1}{2} & \frac{1}{2} & 0 \\ C & \frac{1}{2} & \frac{1}{4} & \frac{1}{4} \\ R & 0 & \frac{1}{2} & \frac{1}{2} \end{array}$$

Find the eigenvalues of T and a formula for T^n. In the long run, what percentage of the days are sunny, cloudy and rainy?

4.8. The eigenvalue method of Section 4.4 for finding general powers of stochastic matrices is only guaranteed to work if the eigenvalues are distinct. Several possibilities occur if the stochastic matrix of a Markov chain has a repeated eigenvalue. The following three examples illustrate these possibilities.

(a) Let

$$T = \begin{bmatrix} \frac{1}{4} & \frac{1}{4} & \frac{1}{2} \\ 1 & 0 & 0 \\ \frac{1}{2} & \frac{1}{4} & \frac{1}{4} \end{bmatrix}$$

be the transition matrix of a three-state Markov chain. Show that T has the repeated eigenvalue $\lambda_1 = \lambda_2 = -\frac{1}{4}$ and $\lambda_3 = 1$, and two distinct eigenvectors

$$\mathbf{r}_1 = \begin{bmatrix} 1 \\ -4 \\ 1 \end{bmatrix} \qquad \mathbf{r}_3 = \begin{bmatrix} 1 \\ 1 \\ 1 \end{bmatrix}.$$

In this case, diagonalization of T is not possible. However it is possible to find a non-singular matrix C such that

$$T = CJC^{-1},$$

where J is the *Jordan decomposition matrix* given by

$$J = \begin{bmatrix} \lambda_1 & 1 & 0 \\ 0 & \lambda_1 & 0 \\ 0 & 0 & 1 \end{bmatrix} = \begin{bmatrix} -\frac{1}{4} & 1 & 0 \\ 0 & -\frac{1}{4} & 0 \\ 0 & 0 & 1 \end{bmatrix},$$

$$C = \begin{bmatrix} \mathbf{r}_1 & \mathbf{r}_2 & \mathbf{r}_3 \end{bmatrix},$$

and $\mathbf{r}_2$ satisfies

$$(T - \lambda_1 I_3)\mathbf{r}_2 = \mathbf{r}_1.$$

Show that we can choose

$$\mathbf{r}_2 = \begin{bmatrix} -10 \\ 24 \\ 0 \end{bmatrix}.$$

Find a formula for J^n and confirm that, as $n \to \infty$,

$$T^n \to \begin{bmatrix} \frac{12}{25} & \frac{1}{5} & \frac{8}{25} \\ \frac{12}{25} & \frac{1}{5} & \frac{8}{25} \\ \frac{12}{25} & \frac{1}{5} & \frac{8}{25} \end{bmatrix}.$$

(b) A four-state Markov chain has the transition matrix

$$S = \begin{bmatrix} 1 & 0 & 0 & 0 \\ \frac{3}{4} & 0 & \frac{1}{4} & 0 \\ 0 & \frac{1}{4} & 0 & \frac{3}{4} \\ 0 & 0 & 0 & 1 \end{bmatrix}.$$

Sketch the transition diagram for the chain, and note that the chain has two absorbing states and is therefore not a regular chain. Show that the eigenvalues of S are $-\frac{1}{4}$, $\frac{1}{4}$ and 1 repeated. Show that there are four distinct eigenvectors. Choose the diagonalizing matrix C as

$$C = \begin{bmatrix} 0 & 0 & -4 & -5 \\ -1 & 1 & -3 & 4 \\ 1 & 1 & 0 & 1 \\ 0 & 0 & 1 & 0 \end{bmatrix}.$$

Find its inverse, and show that, as $n \to \infty$,

$$S^n \to \begin{bmatrix} 1 & 0 & 0 & 0 \\ \frac{4}{5} & 0 & 0 & \frac{1}{5} \\ \frac{1}{5} & 0 & 0 & \frac{4}{5} \\ 0 & 0 & 0 & 1 \end{bmatrix}.$$

Note that since the rows are not the same, this chain does not have an invariant distribution: this is caused by the presence of two absorbing states.

(c) Show that the transition matrix

$$U = \begin{bmatrix} \frac{1}{2} & 0 & \frac{1}{2} \\ \frac{1}{6} & \frac{1}{3} & \frac{1}{2} \\ \frac{1}{6} & 0 & \frac{5}{6} \end{bmatrix}$$

has a repeated eigenvalue, but that, in this case, three independent eigenvectors can be associated with U. Find a diagonalizing matrix C, and find a formula for U^n using $U^n = CD^nC^{-1}$, where

$$D = \begin{bmatrix} \frac{1}{3} & 0 & 0 \\ 0 & \frac{1}{3} & 0 \\ 0 & 0 & 1 \end{bmatrix}.$$

Confirm also that this chain has an invariant distribution.

4.9. Miscellaneous problems on transition matrices. In each case find the eigenvalues of T, a formula for T^n and the limit of T^n as $n \to \infty$. The special cases discussed in Problem 4.8 can occur.

(a)

$$T = \begin{bmatrix} \frac{1}{2} & \frac{7}{32} & \frac{9}{32} \\ 1 & 0 & 0 \\ \frac{1}{2} & \frac{1}{4} & \frac{1}{4} \end{bmatrix};$$

(b)

$$T = \begin{bmatrix} \frac{1}{3} & \frac{1}{4} & \frac{5}{12} \\ 1 & 0 & 0 \\ \frac{1}{4} & \frac{1}{4} & \frac{1}{2} \end{bmatrix};$$

(c)

$$T = \begin{bmatrix} \frac{1}{4} & \frac{3}{16} & \frac{9}{16} \\ \frac{3}{4} & 0 & \frac{1}{4} \\ \frac{1}{2} & \frac{1}{4} & \frac{1}{4} \end{bmatrix};$$

(d)

$$T = \begin{bmatrix} \frac{1}{4} & \frac{1}{4} & \frac{1}{2} \\ \frac{5}{12} & \frac{1}{3} & \frac{1}{4} \\ \frac{1}{2} & \frac{1}{4} & \frac{1}{4} \end{bmatrix};$$

(e)

$$T = \begin{bmatrix} 1 & 0 & 0 & 0 \\ \frac{1}{2} & 0 & 0 & \frac{1}{2} \\ 0 & 0 & 1 & 0 \\ 0 & \frac{1}{2} & \frac{1}{2} & 0 \end{bmatrix}.$$

4.10. A four-state Markov chain has the transition matrix

$$T = \begin{bmatrix} \frac{1}{2} & \frac{1}{2} & 0 & 0 \\ 1 & 0 & 0 & 0 \\ \frac{1}{4} & \frac{1}{2} & 0 & \frac{1}{4} \\ \frac{3}{4} & 0 & \frac{1}{4} & 0 \end{bmatrix}.$$

Find f_i, the probability that the chain returns at some step to state E_i, for each state. Determine which states are transient and which are persistent. Which states form a closed subset? Find the eigenvalues of T, and the limiting behaviour of T^n as $n \to \infty$.

4.11. A six-state Markov chain has the transition matrix

$$T = \begin{bmatrix} \frac{1}{4} & \frac{1}{2} & 0 & 0 & 0 & \frac{1}{4} \\ 0 & 0 & 0 & 0 & 0 & 1 \\ 0 & \frac{1}{4} & 0 & \frac{1}{4} & \frac{1}{2} & 0 \\ 0 & 0 & 0 & 0 & 1 & 0 \\ 0 & 0 & 0 & \frac{1}{2} & \frac{1}{2} & 0 \\ 0 & 0 & 0 & \frac{1}{2} & \frac{1}{2} & 0 \\ 0 & 0 & 1 & 0 & 0 & 0 \end{bmatrix}.$$

Sketch its transition diagram. From the diagram, which states do you think are transient and which do you think are persistent? Which states form a closed subset? Determine the invariant distribution in the subset.

4.12. Draw the transition diagram for the seven-state Markov chain with transition matrix

$$T = \begin{bmatrix} 0 & 1 & 0 & 0 & 0 & 0 & 0 \\ 0 & 0 & 1 & 0 & 0 & 0 & 0 \\ \frac{1}{2} & 0 & 0 & \frac{1}{2} & 0 & 0 & 0 \\ 0 & 0 & 0 & 0 & 1 & 0 & 0 \\ 0 & 0 & 0 & 0 & 0 & 1 & 0 \\ \frac{1}{2} & 0 & 0 & 0 & 0 & 0 & \frac{1}{2} \\ 0 & 0 & 0 & 0 & 0 & 0 & 1 \end{bmatrix}.$$

Hence discuss the periodicity of the states of the chain. From the transition diagram, calculate $p_{11}^{(n)}$ and $p_{44}^{(n)}$ for $n = 2, 3, 4, 5, 6$. (In this example you should confirm that $p_{11}^{(3)} = \frac{1}{2}$ but that $p_{44}^{(3)} = 0$: however, $p_{44}^{(3n)} \neq 0$ for $n = 2, 3, \ldots$ confirming that state E_4 is periodic with period 3.)

4.13. Show that all states in the Markov chain with transition matrix

$$T = \begin{bmatrix} 0 & \frac{3}{4} & \frac{1}{4} \\ \frac{1}{2} & 0 & \frac{1}{2} \\ \frac{3}{4} & \frac{1}{4} & 0 \end{bmatrix}$$

are period 2. Show that $S = T^2$ is the transition matrix of a regular chain. Find its eigenvectors and confirm that S has an invariant distribution given by $\begin{bmatrix} \frac{14}{37} & \frac{13}{37} & \frac{10}{37} \end{bmatrix}$.

4.14. An insect is placed in the maze of cells shown in Figure 4.9. The state E_j is the state in which the insect is in cell j. A transition occurs when the insect moves from one cell to another. Assuming that exits are equally likely to be chosen where there is a choice, construct the transition matrix T for the Markov chain representing the movements of the insect. Show that all states are periodic with period 2. Show that T^2 has two subchains that are

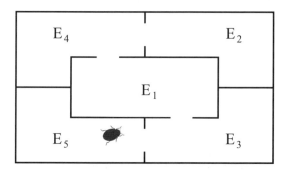

Fig. 4.9 *Insect in a maze.*

both regular. Find the invariant distributions of both subchains. Interpret the results.

4.15. The transition matrix of a four-state Markov chain is given by

$$T = \begin{bmatrix} 1-a & a & 0 & 0 \\ 1-b & 0 & b & 0 \\ 1-c & 0 & 0 & c \\ 1 & 0 & 0 & 0 \end{bmatrix}, \quad (0 < a, b, c < 1).$$

Draw a transition diagram, and, from the diagram, calculate $f_1^{(n)}$, ($n = 1, 2, \ldots$), the probability that a first return to state E_j occurs at the nth step. Calculate also the mean recurrence time μ_1. What type of state is E_1?

4.16. Show that the transition matrix

$$T = \begin{bmatrix} 1-a & a & 0 & 0 \\ 1-a & 0 & a & 0 \\ 1-a & 0 & 0 & a \\ 1 & 0 & 0 & 0 \end{bmatrix}$$

has two imaginary (conjugate) eigenvalues for all a such that $0 < a < 1$. If $a = \frac{1}{2}$, confirm that T has the invariant distribution $\mathbf{p} = \begin{bmatrix} \frac{8}{15} & \frac{4}{15} & \frac{2}{15} & \frac{1}{15} \end{bmatrix}$.

4.17. A production line consists of two manufacturing stages. At the end of each manufacturing stage, each item in the line is inspected, where there is a probability, p, that it will be scrapped, q that it will be sent back to that stage for reworking, and $(1 - p - q)$ that it will be passed to the next stage or completed. The production line can be modelled by a Markov chain with four states: E_1, item scrapped; E_2, item completed; E_3, item in first manufacturing stage; E_4, item in second manufacturing stage. We define states E_1 and E_2 to be absorbing states so that the transition matrix of the chain is

$$T = \begin{bmatrix} 1 & 0 & 0 & 0 \\ 0 & 1 & 0 & 0 \\ p & 0 & q & 1-p-q \\ p & 1-p-q & 0 & q \end{bmatrix}.$$

An item starts along the production line. What is the probability that it is completed in two stages? Calculate $f_3^{(n)}$ and $f_4^{(n)}$. Assuming that $0 < p, q < 1$, what kind of states are E_3 and E_4? What is the probability that an item starting along the production line is ultimately completed?

4.18. Suppose that the production line in the previous problem has N manufacturing stages. What is the probability that an item starting along the production line is ultimately completed?

4.19. The step-dependent transition matrix of Example 4.9 is

$$T_n = \begin{bmatrix} \frac{1}{2} & \frac{1}{2} & 0 \\ 0 & 0 & 1 \\ 1/(n+1) & 0 & n/(n+1) \end{bmatrix}.$$

Find the mean recurrence time for state E_3, and confirm that E_3 is a persistent, non-null state.

4.20. In Example 4.9, a persistent, null state occurred in a chain with step-dependent transitions: such a state cannot occur in a finite chain with a constant transition matrix. However, chains over an infinite number of states can have persistent, null states. Consider the following chain, which has an infinite number of states $E_1, E_2, \ldots$ with the transition probabilities

$$p_{11} = \frac{1}{2}, \quad p_{12} = \frac{1}{2}, \quad p_{j1} = \frac{1}{j+1}, \quad p_{j,j+1} = \frac{j}{j+1}, \quad (j \geq 2).$$

Find the mean recurrence time for E_1, and confirm that E_1 is a persistent, null state.

4.21. A Markov chain has n states and, in the notation of Section 4.7, $p_{jj}^{(n)}$, ($n = 1, 2, \ldots$) is the probability that a return to state E_j occurs at the nth step, and $f_j^{(n)}$ is the probability that a first return occurs at the nth step. It was shown (see (4.28) and (4.29)) that

$$f_j^{(1)} = p_{jj},$$

$$f_j^{(n)} = p_{jj}^{(n)} - \sum_{r=1}^{n-1} f_j^{(r)} p_{jj}^{(n-r)} \qquad (n \geq 2).$$

If

$$H_j(s) = \sum_{n=1}^{\infty} f_j^{(n)} s^n, \qquad G_j(s) = \sum_{n=1}^{\infty} p_{jj}^{(n)} s^n,$$

show that these generating functions satisfy

$$H_j(s)[1 + G_j(s)] = G_j(s).$$

If the state E_j is persistent deduce that $\sum_{n=1}^{\infty} p_{jj}^{(n)}$ is a divergent series.

In Example 4.10, a Markov chain with four states has the transition matrix

$$T = \begin{bmatrix} 0 & \frac{1}{2} & \frac{1}{4} & \frac{1}{4} \\ \frac{1}{2} & \frac{1}{2} & 0 & 0 \\ 0 & 0 & 1 & 0 \\ 0 & 0 & \frac{1}{2} & \frac{1}{2} \end{bmatrix}$$

It was shown in the example that E_1 was a transient state with $f_1 = H_1(1) = \frac{1}{2}$. Show that $G_1(s)$ is a *probability* generating function. Find $H_1(s)$ and $G_1(s)$, and the mean of the distribution $\{p_{11}^{(n)}\}$.

5
Poisson processes

5.1 Introduction

In many applications of stochastic processes, the random variable can be a continuous function of the time t. For example, in a population, births and deaths can occur at any time, and any random variable representing such a probability model must take account of this dependence on time. Other examples include the arrival of telephone calls at an office, or the emission of radioactive particles recorded on a Geiger counter. Interpreting the term *population* in the broad sense (not simply individuals, but particles, telephone calls, etc depending on the context) we might be interested typically in the probability that the population size is, say, n at time t. We shall represent this probability usually by $p_n(t)$. For the Geiger counter application it will represent the probability that n particles have been recorded up to time t, whilst for the arrival of telephone calls it will represent the number of calls logged up to time t.

5.2 The Poisson process

Let $N(t)$ be a time-varying random variable representing the population size at time t. Consider the continuous-time probability distribution given by

$$p_n(t) = \mathbf{P}[N(t) = n] = \frac{(\lambda t)^n e^{-\lambda t}}{n!}, \qquad (t \geq 0) \tag{5.1}$$

for $n = 0, 1, 2, \dots$. It is assumed that $N(t)$ can take the integer values $n = 0, 1, 2, \dots$. Since (5.1) is a probability distribution we observe that

$$\sum_{n=0}^{\infty} p_n(t) = \sum_{n=0}^{\infty} \frac{(\lambda t)^n}{n!} e^{-\lambda t} = 1,$$

using the power series expansion for the exponential function.

In fact (see Section 1.7) $p_n(t)$ is a *Poisson distribution* with parameter λt. For this reason any application for which (5.1) holds is known as a *Poisson process*: we shall give a full formal statement of the process in Section 5.8. Some sample distributions for $p_n(t)$ versus λt in the cases $n = 0, 1, 2, 3$ are shown in Figure 5.1. The Poisson process is a special case of the more general birth process, which is developed in Chapter 6.

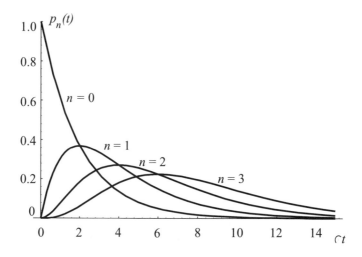

Fig. 5.1 *Probabilities $p_n(t)$ for $\lambda = 0.5$ and $n = 0, 1, 2, 3$.*

Since, for $n \geq 1$,

$$\frac{dp_n(t)}{dt} = \frac{(n - \lambda t)}{n!}\lambda^n t^{n-1}e^{-\lambda t}, \tag{5.2}$$

the maximum values of the probabilities for fixed n occur at $t = n/\lambda$, where $dp_n(t)/dt = 0$. For $n = 0$,

$$\frac{dp_0(t)}{dt} = -\lambda e^{-\lambda t}. \tag{5.3}$$

The mean $\mu(t)$ of the Poisson distribution will be a function of time given by the sum of the products of the possible outcomes n and their probable occurrences $p_n(t)$, namely,

$$\mu(t) = \mathbf{E}[N(t)] = \sum_{n=1}^{\infty} np_n(t) = \sum_{n=1}^{\infty} n\frac{(\lambda t)^n}{n!}e^{-\lambda t}$$

$$= e^{-\lambda t}\lambda t \sum_{n=1}^{\infty} \frac{(\lambda t)^{n-1}}{(n-1)!} = e^{-\lambda t}\lambda t e^{\lambda t} = \lambda t. \tag{5.4}$$

Note that the mean value increases linearly with time.

This observation that the mean increases steadily with time gives a pointer as to why the Poisson distribution is appropriate for cumulative recording, such as the Geiger counter application and the call-logging problem for the telephone operator. For the Geiger counter, the recording of radioactive particle hits is assumed

to occur at random with the probability of a new subsequent hit being independent of any previous recordings. This independence requirement is crucial for
the Poisson process. It seems a reasonable assumption for a radioactive test of a
sufficiently large amount of material, in which the number of hits over a period of
time is minute compared with the total number of radioactive particles. We can
show how this arises from the Poisson process as follows.

Example 5.1. Find the variance of the Poisson distribution $p_n(t)$ $(n=0, 1, 2, \ldots)$.

In terms of means, the variance is given by (see Section 1.6)

$$\mathbf{V}[N(t)] = \mathbf{E}[N(t)^2] - [\mathbf{E}[N(t)]]^2 = \mathbf{E}[N(t)^2] - (\lambda t)^2$$

by (5.4). Also

$$\mathbf{E}[N(t)^2] = \sum_{n=1}^{\infty} \frac{n^2 (\lambda t)^n}{n!} e^{-\lambda t}$$

$$= e^{-\lambda t} \sum_{n=1}^{\infty} \frac{n (\lambda t)^n}{(n-1)!}$$

$$= e^{-\lambda t} t \frac{d}{dt} \sum_{n=1}^{\infty} \frac{\lambda^n t^n}{(n-1)!}$$

$$= e^{-\lambda t} t \frac{d}{dt} [\lambda t e^{\lambda t}] = \lambda t + (\lambda t)^2.$$

Thus

$$\mathbf{V}[N(t)] = \lambda t + (\lambda t)^2 - (\lambda t)^2 = \lambda t.$$

Note that the Poisson distribution has the property that its mean is the same as its
variance (see Section 1.7). □

From equations (5.1), (5.2) and (5.3) it follows that

$$\frac{dp_0(t)}{dt} = -\lambda p_0(t), \tag{5.5}$$

$$\frac{dp_n(t)}{dt} = \lambda [p_{n-1}(t) - p_n(t)], \qquad (n \geq 1). \tag{5.6}$$

These are *differential-difference equations* for the sequence of probabilities $p_n(t)$.
From the definition of differentiation, the derivatives are obtained by the limiting process

$$\frac{dp_n(t)}{dt} = \lim_{\delta t \to 0} \frac{p_n(t + \delta t) - p_n(t)}{\delta t},$$

so that approximately, for small $\delta t > 0$,

$$\frac{dp_n(t)}{dt} \approx \frac{p_n(t + \delta t) - p_n(t)}{\delta t}.$$

Thus, eliminating the derivatives in (5.5) and (5.6) in favour of their approximations, we can replace the equations by

$$\frac{p_0(t + \delta t) - p_0(t)}{\delta t} \approx -\lambda p_0(t),$$

$$\frac{p_n(t + \delta t) - p_n(t)}{\delta t} \approx \lambda [p_{n-1}(t) - p_n(t)],$$

so that

$$\left. \begin{aligned} p_0(t) &\approx (1 - \lambda \delta t) p_0(t) \\ p_n(t + \delta t) &\approx p_{n-1}(t)\lambda \delta t + p_n(t)(1 - \lambda \delta t), \qquad (n \geq 1) \end{aligned} \right\} \qquad (5.7)$$

We can interpret the equations as follows. We can infer from these formulas that the probability that a particle is recorded in the short time interval δt is $\lambda \delta t$, and that the probability that two or more particles are recorded is negligible, and consequently that no recording takes place with probability $(1 - \lambda \delta t)$. In (5.7), the only way in which the outcome reading $n(\geq 1)$ can occur at time $t + \delta t$ is that either one particle was recorded in the interval δt when n particles were recorded at time t, or that nothing occurred with probability $(1 - \lambda \delta t)$ when n particles were recorded at time t. In fact, (5.7) is really a re-statement of the *partition theorem* or *law of total probability*, and is often the starting point for the modelling of random processes in continuous time. We will look at this approach for the Poisson process in the next section, before we develop it further in the next chapter for birth and death processes.

5.3 Partition theorem approach

We can use equation (5.7) as the starting approach using the partition theorem (Section 1.3). We argue as in the last paragraph of the previous section but tighten the argument. For the Geiger counter application (a similar argument can be adapted for the call-logging problem, etc) we assume that the probability that one particle is recorded in the short time interval δt is

$$\lambda \delta t + o(\delta t).$$

(The term $o(\delta t)$ described in words as 'little o δt' means that the remainder or error is *of lower order than* δt, that is,

$$\lim_{\delta t \to 0} \frac{o(\delta t)}{\delta t} = 0.$$

The probability of two or more hits is assumed to be $o(\delta t)$, that is, negligible as $\delta t \to 0$, and the probability of no hits is $1 - \lambda \delta t + o(\delta t)$. We now apply the partition theorem on the possible outcomes. The case $n = 0$ is special since reading zero can only occur through no event occurring. Thus

$$p_0(t + \delta t) = [1 - \lambda \delta t + o(\delta t)]p_0(t),$$

$$p_n(t+\delta t) = p_{n-1}(t)(\lambda\delta t + o(\delta t)) + p_n(t)(1 - \lambda\delta t + o(\delta t)) + o(\delta t) \qquad (n \geq 1).$$

Dividing through by δt and re-organizing the equations, we find that

$$\frac{p_0(t + \delta t) - p_0(t)}{\delta t} = -\lambda p_0(t) + o(1),$$

$$\frac{p_n(t + \delta t) - p_n(t)}{\delta t} = \lambda[p_{n-1}(t) - p_n(t)] + o(1)$$

Now let $\delta t \to 0$. Then, by the definition of the derivative,

$$\frac{dp_0(t)}{dt} = -\lambda p_0(t), \qquad\qquad (5.8)$$

$$\frac{dp_n(t)}{dt} = \lambda[p_{n-1}(t) - p_n(t)], \qquad (n \geq 1) \qquad (5.9)$$

All terms $o(1)$ tend to zero as $\delta t \to 0$. Not surprisingly we have recovered equations (5.5) and (5.6).

We started this chapter by looking at the Poisson distributions defined by (5.1). We now look at techniques for solving (5.8) and (5.9), although for the Poisson process we know the solutions. The methods give some insight into the solution of more general continuous-time random processes.

5.4   Iterative method

Suppose that our model of the Geiger counter is based on equations (5.8) and (5.9), and that we wish to solve these equations to recover $p_n(t)$, which we assume to be unknown for the purposes of this exercise. Equation (5.8) is an ordinary differential equation for one unknown function $p_0(t)$. It is of first-order, and it can be easily verified that its general solution is

$$p_0(t) = C_0 e^{-\lambda t}, \qquad\qquad (5.10)$$

where C_0 is a constant. We need to specify *initial conditions* for the problem. Assume that the instrumentation of the Geiger counter is set to zero initially. Thus, we have the *certain* event for which $p_0(0) = 1$, and consequently $p_n(0) = 0$, $(n \geq 1)$: the probability of any reading other than zero is zero at time $t = 0$. Hence $C_0 = 1$ and

$$p_0(t) = e^{\lambda t}. \qquad\qquad (5.11)$$

Now put $n = 1$ in (5.9) so that

$$\frac{dp_1(t)}{dt} = \lambda p_0(t) - \lambda p_1(t),$$

or

$$\frac{dp_1(t)}{dt} + \lambda p_1(t) = \lambda p_0(t) = \lambda e^{-\lambda t},$$

after substituting for $p_0(t)$ from equation (5.11). This first-order differential equation for $p_1(t)$ is of *integrating factor* type with integrating factor

$$e^{\int \lambda \, dt} = e^{\lambda t},$$

in which case it can be rewritten as the following separable equation:

$$\frac{d}{dt}\left(e^{\lambda t} p_1(t)\right) = \lambda e^{\lambda t} e^{-\lambda t} = \lambda.$$

Hence, integration with respect to t results in

$$e^{\lambda t} p_1(t) = \int \lambda \, dt = \lambda t + C_1.$$

Thus

$$p_1(t) = \lambda t e^{-\lambda t} + C_1 e^{-\lambda t} = \lambda t e^{-\lambda t}$$

since $p_1(0) = 0$. We now repeat the process by putting $n = 2$ in (5.9) and substituting in the $p_1(t)$ which has just been found. The result is the equation

$$\frac{dp_2(t)}{dt} + \lambda p_2(t) = \lambda p_1(t) = \lambda t e^{-\lambda t}.$$

This is a further first-order integrating-factor differential equation, which can be solved using the same method. The result is, using the initial condition $p_2(0) = 0$,

$$p_2(t) = \frac{(\lambda t)^2 e^{-\lambda t}}{2!}.$$

The method can be repeated for $n = 3, 4, \ldots$, and the results imply that

$$p_n(t) = \frac{(\lambda t)^n e^{-\lambda t}}{n!}, \tag{5.12}$$

which is the probability given by (5.1) as we would expect. The result in (5.12) can be justified rigorously by constructing a proof by induction.

The initial condition is built into the definition of the Poisson process (see the summary in Section 5.8). The iterative approach outlined above does permit the use of other initial conditions such as the assumption that the Geiger counter is set to reading n_0 at time $t = 0$ (see Problem 5.8).

The iterative method works for the differential-difference equations (5.9) because they contain forward differencing only in n. In many applications, both forward and backward differencing appear in the equations, with the result that successive methods of solution can no longer be applied. The alternative generating function approach will be explained in the next section.

5.5   The generating function

An alternative approach to the solution of differential difference equations uses the *probability generating function* first introduced in Section 1.9.   For continuous-time random processes, we define a generating function as a power series in a dummy variable s, say, in which the coefficients in the series are the probabilities, the $p_n(t)$s in our notation here.   Thus, we construct a generating function $G(s, t)$ as

$$G(s, t) = \sum_{n=0}^{\infty} p_n(t)s^n. \tag{5.13}$$

Here, s is a dummy variable, which is in itself not of much direct interest. However, given the function $G(s, t)$, $p_n(t)$ can be recovered from the series obtained by expanding $G(s, t)$ in powers of s, and looking at their coefficients.

The practical steps involved in expressing the differential-difference equations in terms of the generating function are as follows. Multiply equation (5.9) by s^n, and add the equations together including (5.8). In summation notation the result is

$$\sum_{n=0}^{\infty} \frac{\mathrm{d}p_n(t)}{\mathrm{d}t}s^n = \lambda \sum_{n=1}^{\infty} p_{n-1}(t)s^n - \lambda \sum_{n=0}^{\infty} p_n(t)s^n. \tag{5.14}$$

Note carefully the lower limits on the summations. Note also that the right-hand side of (5.8) has been included in the second series on the right-hand side of (5.14). We attempt to express each of the series in (5.14) in terms of the generating function $G(s, t)$ or its partial derivatives with respect to either s or t. Thus, looking at each of the series in turn, we find that

$$\sum_{n=0}^{\infty} \frac{\mathrm{d}p_n(t)}{\mathrm{d}t}s^n = \frac{\partial}{\partial t} \sum_{n=0}^{\infty} p_n(t)s^n = \frac{\partial G(s, t)}{\partial t},$$

$$\sum_{n=1}^{\infty} p_{n-1}(t)s^n = \sum_{m=0}^{\infty} p_m(t)s^{m+1} = sG(s, t), \quad \text{(putting } n = m + 1\text{)},$$

$$\sum_{n=0}^{\infty} p_n(t)s^n = G(s, t).$$

We can now replace (5.14) by the equation

$$\frac{\partial G(s, t)}{\partial t} = \lambda s G(s, t) - \lambda G(s, t) = \lambda(s - 1)G(s, t). \tag{5.15}$$

This *partial* differential equation in $G(s, t)$ replaces the set of differential-difference equations (5.8) and (5.9).

If (5.15) can be solved for the generating function $G(s, t)$, and if this function can then be expanded in powers of s, then the probabilities can be read off from this expansion.

Since equation (5.15) only contains a derivative with respect to t, it behaves more like an ordinary differential equation. We can integrate it with respect to t (it is a similar equation to (5.8)) so that

$$G(s, t) = A(s)e^{\lambda(s-1)t}, \tag{5.16}$$

where $A(s)$ is the 'constant of integration' but nevertheless can be a function of the other variable s. That (5.16) is the solution can be verified by direct substitution in (5.15).

The function $A(s)$ is determined by the initial conditions that must now be expressed in terms of the generating function. The initial conditions $p_0(0) = 1$, $p_n(0) = 0$, $(n \geq 1)$ translate into

$$G(s, 0) = \sum_{n=0}^{\infty} p_n(0)s^n = 1.$$

Generally, the initial conditions lead to $G(s, 0)$ being a specified function in s. Thus, in (5.16), $A(s) = 1$ and the required generating function for the Poisson process is

$$G(s, t) = e^{\lambda(s-1)t}. \tag{5.17}$$

To obtain the individual probabilities, we expand the generating function in powers of s. In this case we need the power series for the exponential function, which is

$$e^z = 1 + z + \frac{z^2}{2!} + \frac{z^3}{3!} + \cdots = \sum_{n=0}^{\infty} \frac{z^n}{n!}.$$

Applying this result to $G(s, t)$, we obtain

$$G(s, t) = e^{\lambda(s-1)t} = e^{-\lambda t}e^{\lambda st} = e^{-\lambda t}\sum_{n=0}^{\infty} \frac{(\lambda s t)^n}{n!}$$

$$= \sum_{n=0}^{\infty} \frac{(\lambda t)^n e^{\lambda t}}{n!}s^n.$$

Hence, the probability $p_n(t)$ is confirmed again as

$$p_n(t) = \frac{(\lambda t)^n e^{-\lambda t}}{n!}.$$

The mean at time t can also be expressed in terms of the generating function. Quite generally, the mean is given by (see also Section 1.9)

$$\mu(t) = \sum_{n=1}^{\infty} np_n(t) = \left.\frac{\partial G(s, t)}{\partial s}\right|_{s=1} = G_s(1, t). \tag{5.18}$$

For the Poisson process above

$$\mu(t) = \left.\frac{\partial}{\partial s}e^{\lambda(s-1)t}\right|_{s=1} = \left.\lambda t e^{\lambda(s-1)t}\right|_{s=1} = \lambda t,$$

which confirms (5.4) again.

5.6   Variance for the Poisson process

In a similar manner, the variance of a process can be represented by a formula in terms of $G(s, t)$: this is a simple extension of the result in Section 1.9 to $G(s, t)$. From Section 1.6, the variance of a random variable X is defined as

$$\sigma^2 = \mathbf{V}(X) = \mathbf{E}(X^2) - [\mathbf{E}(X)]^2 = \sum_{n=1}^{\infty} n^2 p_n(t) - \mu(t)^2, \qquad (5.19)$$

if the probability distribution is $\{p_n(t)\}$, $(n = 0, 1, 2, \ldots)$. As we saw in the previous section

$$\mu(t) = G_s(1, t).$$

Differentiating $G(s, t)$ twice with respect to s, we have

$$\frac{\partial^2}{\partial s^2} G(s, t) = G_{ss}(s, t) = \frac{\partial^2}{\partial s^2} \sum_{n=0}^{\infty} p_n(t)s^n = \sum_{n=2}^{\infty} n(n-1)p_n(t)s^{n-2}$$

$$= \sum_{n=2}^{\infty} p_n(t)s^{n-2} - \sum_{n=2}^{\infty} n p_n(t)s^{n-2}.$$

Now put $s = 1$ in this derivative:

$$G_{ss}(1, t) = \sum_{n=2}^{\infty} n^2 p_n(t) - G_s(1, t) + p_1(t) = \sum_{n=1}^{\infty} n^2 p_n(t) - G_s(1, t). \quad (5.20)$$

From (5.19) and (5.20) we obtain the general formula

$$\sigma^2 = \mathbf{V}(X) = G_{ss}(1, t) + G_s(1, t) - G_s(1, t)^2. \qquad (5.21)$$

Example 5.2. Find the variance of the Poisson process with parameter λt.

The probability generating function of the Poisson process is $G(s, t) = e^{\lambda(s-1)t}$. From the previous section

$$\mu(t) = G_s(1, t) = \lambda t.$$

Also

$$\frac{\partial^2}{\partial s^2} G(s, t) = \lambda^2 t^2 e^{\lambda(s-1)t}.$$

Hence from (5.21), the variance is given by

$$\mathbf{V}(X) = \lambda^2 t^2 + \lambda t - \lambda^2 t^2 = \lambda t,$$

which is the same as the mean for the Poisson process. This confirms the result obtained directly in Example 5.1. $\qquad\qquad \square$

5.7 Arrival times

For the Geiger counter, the *arrival time* T_n for reading n is defined as the earliest time at which the random variable $X_t = n$. In particular, $T_0 = 0$. Figure 5.2 shows how the times occur in a Poisson process. The *inter-arrival time* $Q_n = T_n - T_{n-1}$ is the time between successive hits. We have assumed that the particles hit the Geiger counter randomly and independently. Hence, we might reasonably expect that the random variable giving the arrival time of the first particle will have the same distribution as the inter-arrival times between any two successive readings. It is as if the counter has reset itself to zero and is awaiting the arrival of the next particle; that is, it has no memory or Markov property (see Problem 5.2). At time t the probability that no particle has been detected by the Geiger counter is $p_0(t) = e^{-\lambda t}$. The distribution function of the arrival times of the *next* reading is therefore given by

$$F(s) = 1 - p_0(s) = 1 - e^{-\lambda s},$$

which leads to the exponential probability density function (see Section 1.8)

$$f(s) = \frac{dF(s)}{ds} = \lambda e^{-\lambda s}.$$

The mean or expected value of the inter-arrival times is given by

$$\mu = \int_0^\infty \lambda s e^{-\lambda s}\, ds$$

$$= \left[-se^{-\lambda s} \right]_0^\infty + \int_0^\infty e^{-\lambda s}\, ds, \qquad \text{(integrating by parts)}$$

$$= 0 + \left[-\frac{e^{-\lambda s}}{\lambda} \right]_0^\infty = \frac{1}{\lambda}.$$

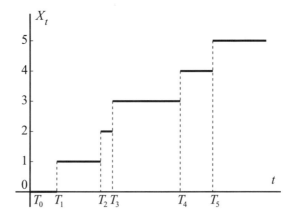

Fig. 5.2 *The Poisson process and the arrival times $T_0, T_1, T_2, \ldots$.*

Hence, the mean of the inter-arrival times or the mean of the times between successive readings for the Geiger counter is the reciprocal of the parameter λ.

Example 5.3. Incoming telephone calls to an operator are assumed to be a Poisson process with parameter λ. Find the density function of the length of time for n calls to be received and find the mean time and variance of the random variable of the length of time for n calls.

We are now interested in the time T_n, which is the earliest time at which the random variable $N(t) = n$, and its distribution. The probability distribution of the random variable T_n is given by

$$
\begin{aligned}
F(t) &= \mathbf{P}\{T_n \le t\} \\
&= \mathbf{P}\{n \text{ or more calls have arrived in the time interval } (0, t)\} \\
&= p_n(t) + p_{n+1}(t) + \cdots \\
&= \sum_{r=n}^{\infty} p_r(t) = \sum_{r=n}^{\infty} \frac{(\lambda t)^r e^{-\lambda t}}{r!}
\end{aligned}
$$

using (5.1). The corresponding density function is

$$
f(t) = \frac{\mathrm{d}F(t)}{\mathrm{d}t} = \sum_{r=n}^{\infty} \left[\frac{\lambda(\lambda t)^{r-1} e^{-\lambda t}}{(r-1)!} - \frac{\lambda(\lambda t)^r e^{-\lambda t}}{r!} \right] = \frac{\lambda(\lambda t)^{n-1}}{(n-1)!} e^{-\lambda t},
$$

which is the density function of a gamma distribution with parameters n and λ. The mean of this density function is

$$
\mu = \mathbf{E}(T_n) = \int_0^{\infty} t f(t)\mathrm{d}t = \int_0^{\infty} \frac{(\lambda t)^n e^{-\lambda t}}{(n-1)!}\mathrm{d}t = \frac{1}{\lambda(n-1)!} \int_0^{\infty} u^n e^{-u}\mathrm{d}u = \frac{n}{\lambda},
$$

(see Section 1.8).

The variance of this gamma distribution is given by (see Section 1.8, again)

$$
\begin{aligned}
\mathbf{V}(T_n) = \mathbf{E}(T_n^2) - [\mathbf{E}(T_n)]^2 &= \int_0^{\infty} \frac{t^2 \lambda(\lambda t)^{n-1}}{(n-1)!} e^{\lambda t}\mathrm{d}t - \mu^2 \\
&= \frac{1}{\lambda(n-1)!} \int_0^{\infty} (\lambda t)^{n+1} e^{-\lambda t}\mathrm{d}t - \frac{n^2}{\lambda^2} \\
&= \frac{1}{\lambda^2(n-1)!} \int_0^{\infty} u^{n+1} e^{-u}\mathrm{d}u - \frac{n^2}{\lambda^2} \\
&= \frac{(n+1)!}{\lambda^2(n-1)!} - \frac{n^2}{\lambda^2} = \frac{n(n+1)}{\lambda^2} - \frac{n^2}{\lambda^2} = \frac{n}{\lambda^2}.
\end{aligned}
$$

$\square$

Example 5.4. A fire and emergency rescue service receives calls for assistance at a rate of 10 per day. Teams man the service in 12 hour shifts. Assume that requests for help form a Poisson process.

(i) From the beginning of a shift how long would the team expect to wait until their first call?

(ii) What is the probability that a team would receive 6 requests for help in a shift?

(iii) What is the probability that a team has no requests for assistance in a shift?

(iv) Of calls for assistance, one in five is a false alarm. What is the probability that a team has 6 requests for help in a shift but no false alarms?

Let time be measured in hours. Then 10 calls per day is equivalent to 10/24 per hour. Hence, the Poisson rate is $\lambda = 0.4167$. In this problem, $p_n(t) = (\lambda t)^n e^{-\lambda t}/n!$ is the probability that there are n calls in time t, where t is measured in hours.

(i) From equation (5.19) the team would expect to wait for $1/\lambda = 2.4$ hours until the first emergency.

(ii) Using equation (5.12), the required probability is

$$\mathbf{P}[N(12) = 6] = p_6(12) = \frac{1}{6!}\left[\frac{10}{24}.12\right]^6 e^{-\frac{10}{24}.12} = 0.146.$$

(iii) The probability of no request will be

$$\mathbf{P}[N(12) = 0] = p_0(12) = e^{-\frac{10}{24}.12} = e^{-5} = 0.00674.$$

(iv) The genuine calls will be Poisson with rate $\lambda' = 8/24$, say. If $N'(t)$ is the random variable of the genuine calls with distribution $p'_n(t)$, then there will be six genuine calls in a shift with probability

$$\mathbf{P}\{X'_{12} = 6\} = p'_6(12) = \frac{4^6}{6!}e^{-4} = 0.104.$$

□

5.8 Summary of the Poisson process

We have introduced the Poisson process in an informal way, mainly through the illustrative model application of the Geiger counter. We can formally summarize what is meant by a Poisson process as follows.

Consider the random process with the random variable $N(t)$, ($t \geq 0$). Then $N(t)$ is a random variable of a Poisson process if

(i) $N(t)$ can take the values $\{0, 1, 2, \ldots\}$;

(ii) $N(0) = 0$;

(iii) $N(t_1) \leq N(t_2)$ if $t_2 \geq t_1$;

(iv) for any sequence $0 < t_1 < t_2 \cdots < t_n$, the random variables $N(t_i) - N(t_{i-1})$ (number of events in non-overlapping time intervals) are mutually independent;

(v) $\mathbf{P}\{X_{t+\delta t} = n+1 | X_t = n\} = \lambda \delta t + o(\delta t)$,
$\mathbf{P}\{X_{t+\delta t} \geq n+2 | X_t = n\} = o(\delta t)$;

(vi) the probability generating function is

$$G(s,t) = e^{\lambda(s-1)t},$$

subject to $p_0(0) = 1$;

(vii) the inter-arrival times are exponentially distributed and independent.

In (v) the first conditional probability specifies that the probability of an event in the time interval $(t, t+h)$ behaves linearly in h for h small, whilst the second conditional probability states that the likelihood of two or more events taking place in this time interval is negligible.

Problems

5.1. A Geiger counter is used to investigate a sample that contains two radioactive materials. Assuming that the readings for each material separately are Poisson processes with rates λ_1 and λ_2, show that the mixture behaves as a Poisson process with rate $\lambda_1 + \lambda_2$.

5.2. The probability that no event has taken place in time s after the previous event in a Poisson process with rate λ has the exponential distribution given by $e^{-\lambda s}$. In other words, if Q_n represents this random variable, then

$$\mathbf{P}\{Q_n > s\} = e^{-\lambda s}.$$

Show that, if $s_1, s_2 \geq 0$, then

$$\mathbf{P}\{Q_n > s_1 + s_2 | Q_n > s_1\} = \mathbf{P}\{Q_n > s_2\}.$$

What does this result imply about the Poisson process and its memory of past events?

5.3. A random variable N_t has the distribution

$$p_n(t) = \frac{(\lambda t)^{n-1}}{(n-1)!} e^{-\lambda t}, \qquad (n = 1, 2, 3, \ldots).$$

Find its mean and variance.

5.4. The variance of a random variable X_t is given by

$$\mathbf{V}(X_t) = \mathbf{E}(X_t^2) - \mathbf{E}(X_t)^2.$$

In terms of the generating function $G(s,t)$, show that

$$\mathbf{V}(X_t) = \left[\frac{\partial}{\partial s}\left(s \frac{\partial G(s,t)}{\partial s} \right) - \left(\frac{\partial G(s,t)}{\partial s} \right)^2 \right]_{s=1}.$$

(an alternative formula to (5.20)). Obtain the variance for the Poisson process using the generating function

$$G(s, t) = e^{\lambda(s-1)t}$$

given by equation (5.17), and check your answer with that given in Problem 5.3.

5.5. Explain why any probability generating function $G(s, t)$ must satisfy $G(1, t) = 1$. What do $G(1, 0)$ and $G(0, t)$ represent?

5.6. A telephone answering service receives calls whose frequency varies with time but independently of other calls, perhaps with a daily pattern – more during the day than the night. The rate $\lambda(t) \geq 0$ becomes a function of the time t. The probability that a call arrives in the small time interval $(t, t+\delta t)$ when n calls have been received at time t satisfies

$$p_n(t+\delta t) = p_{n-1}(t)(\lambda(t)\delta t + o(\delta t)) + p_n(t)(1 - \lambda(t)\delta t + o(\delta t)), \quad (n \geq 1),$$

with

$$p_0(t) = (1 - \lambda(t)\delta t + o(\delta t))p_0(t).$$

It is assumed that the probability of two or more calls arriving in the interval $(t, t + \delta t)$ is negligible. Find the set of differential-difference equations for $p_n(t)$. Obtain the probability generating function $G(s, t)$ for the process and confirm that it is a Poisson process with parameter $\int_0^t \lambda(s)ds$. Find $p_n(t)$ by expanding $G(s, t)$ in powers of s. What is the mean number of calls received up to time t?

5.7. For the telephone answering service in Problem 5.6, suppose that the rate is periodic and given by $\lambda(t) = a + b\cos(\omega t)$ where $a > 0$ and $|b| < a$. Using the probability generating function from Problem 5.6 find the probability that n calls have been received by time t. Find also the mean number of calls received by time t. Sketch graphs of $p_0(t)$, $p_1(t)$ and $p_2(t)$ where $a = 0.5$, $b = 0.2$ and $\omega = 1$.

5.8. A Geiger counter is pre-set so that its initial reading is n_0 at time $t = 0$. What are the initial conditions on $p_n(t)$, the probability that the reading is n at time t, and its generating function $G(s, t)$? Find $p_n(t)$, and the mean reading of the counter at time t.

5.9. A Poisson process with probabilities

$$p_n(t) = \mathbf{P}[N(t) = n] = \frac{(\lambda t)^n e^{-\lambda t}}{n!}$$

has a random variable $N(t)$. Calculate the following probabilities associated in the process:

(a) $\mathbf{P}[N(3) = 6]$;

(b) $\mathbf{P}[N(2.6) = 3]$;

(c) $\mathbf{P}[N(3.7) = 4|N(2.1) = 2]$;

(d) $\mathbf{P}[N(7) - N(3) = 2|N(3) = 2]$;

(e) $\mathbf{P}[N(7) - N(3) = 3]$.

5.10. A telephone banking service receives an average of 1000 call per hour. On average, a customer transaction takes one minute. If the calls arrive as a Poisson process, how many operators should the bank employ to avoid accumulation of incoming calls?

5.11. A Geiger counter automatically switches off when the nth particle has been recorded where n is fixed. The arrival of recorded particles is assumed to be a Poisson process with parameter λt. What is the expected value of the switch-off times?

5.12. Particles are emitted from a radioactive source, and N_t, the random variable of the number of particles emitted up to time t from $t = 0$, is a Poisson process with parameter λt. The probability that any particle hits a certain target is p, independently of any other particle. If M_t is the random variable of the number of particles that hit the target up to time t, show, using the law of total probability, that M_t forms a Poisson process with parameter $\lambda p t$.

6
Birth and death processes

6.1 Introduction

We shall now continue our investigation of further random processes in continuous time. In the Poisson process in Chapter 5, the probability of a further event was independent of the current number of events, or readings in the Geiger counter analogy: this was a specific assumption in the definition of the Poisson process (Section 4.7). On the other hand, in birth and death processes, the probability of a birth or death will depend on the population size at time t. The more individuals in the population, the greater the possibility of a death for example. Birth and death processes and Poisson processes form part of a wider class of random processes that are often referred to as *Markov processes*. Additionally, they include queueing processes, epidemics, predator–prey competition and others. Markov processes are characterized by the condition that future development of these processes depends only on their current states and not their history up to that time. Generally, Markov processes are easier to model and analyse, and they do include many interesting applications. Non-Markov processes, in which the future state of a process depends on its whole history, are generally harder to analyse mathematically.

We have adopted a gradual approach to the full problem. The birth and death processes are looked at separately, and then combined into the full birth and death process. Generally, the partition theorem approach is used to derive the equations for $p_n(t)$, the probability that the population size is n at time t, and the probability generating function approach is the preferred method of solution.

6.2 The birth process

In this process, everyone lives for ever: there are no deaths. This process could model a bacterium in which each cell randomly and independently divides into two cells at some future time, and the same happens for each divided cell. The births could start with n_0 cells at time $t = 0$. We shall assume in this first model that the probability that any individual cell divides in the time interval $(t, t + \delta t)$ is proportional to δt for small δt. If λ is the birth-rate associated with this process, then the probability that the cell divides is $\lambda \delta t$ in the interval, and this is mutually exclusive from any other event. The probability that n cells divide is simply $\lambda n \delta t$. To avoid complications, we assume that the probability that two or more births take place in the time interval δt is $o(\delta t)$ (that is, it can be ignored) with the consequence that the probability that no cell divides is $1 - \lambda n \delta t - o(\delta t)$.

There are many assumptions in this model, including not allowing multiple births (twins, triplets, etc) which may be significant in any real situation. The probability that a cell divides may not be homogeneous in time – it could decline with time, for example; the probability could also depend both on the state of the host and the number of cells, in which case the parameter λ could be a function of the population size n. However, the *simple* birth process described above is an interesting starting point for studying a stochastic model of growth. It is also known as the *Yule process* named after one of its originators.

This birth process is an example of a continuous-time Markov chain (see Chapter 4) with an unbounded set of states $E_{n_0}, E_{n_0+1}, E_{n_0+2}, \ldots$, where E_n is the state in which the population size is n. In this chain, however, transitions can only take place between n and $n + 1$ since there are no deaths. The probability that a transition occurs from E_n to E_{n+1} in time δt is $\lambda n \delta t$ and that no transition occurs is $1 - \lambda n \delta t$. In a continuous-time Markov chain, the chain may spend varying periods of time in any state as the population grows.

If N_t is the random variable associated with the process then we write

$$\mathbf{P}\{N(t) = n\} = p_n(t),$$

where $p_n(t)$ is the probability that the population size is n at time t. If the initial population size is $n_0 \geq 1$ at time $t = 0$, then

$$p_{n_0}(0) = 1, \text{ and } p_n(0) = 0 \text{ for } n \neq n_0. \tag{6.1}$$

According to the rules outlined above, a population of size n at time $t + \delta t$ can arise either from a population of size $n - 1$ at time t with a birth occurring with probability $\lambda(n - 1)\delta t + o(\delta t)$ or through no event in a population of size n which can occur with probability

$$p_{n_0}(t + \delta t) = p_{n_0}(t)[1 - \lambda n_0 \delta t + o(\delta t)],$$

$$p_n(t+\delta t) = p_{n-1}(t)[\lambda(n-1)\delta t+o(\delta t)]+p_n(t)[1-\lambda n\delta t+o(\delta t)] \qquad (n \geq n_0+1)$$

which can be re-arranged into

$$\frac{p_{n_0}(t + \delta t) - p_{n_0}(t)}{\delta t} = -\lambda n_0 p_{n_0}(t) + o(1),$$

$$\frac{p_n(t + \delta t) - p_n(t)}{\delta t} = \lambda(n - 1)p_{n-1}(t) - \lambda n p_n(t) + o(1).$$

Let $\delta t \to 0$ with the result that

$$\left.\begin{aligned}\frac{\mathrm{d}p_{n_0}(t)}{\mathrm{d}t} &= -\lambda n p_{n_0}(t),\\[1mm]\frac{\mathrm{d}p_n(t)}{\mathrm{d}t} &= \lambda(n-1)p_{n-1}(t) - \lambda n p_n(t), \quad (n \geq n_0 + 1)\end{aligned}\right\} \tag{6.2}$$

This is the differential-difference equation for the simple birth process. Since this is a *birth* process it follows that $p_n(t) = 0$ for $n < n_0$.

The system of equations (6.2) can be solved successively starting with $n = n_0$. The first equation

$$\frac{\mathrm{d}p_{n_0}(t)}{\mathrm{d}t} = -\lambda n_0 p_{n_0}(t)$$

has the solution

$$p_{n_0}(t) = \mathrm{e}^{-\lambda n_0 t},$$

since $p_{n_0}(0) = 1$. Put $n = n_0 + 1$ in (6.2). Then the next equation is

$$\frac{\mathrm{d}p_{n_0+1}(t)}{\mathrm{d}t} - \lambda(n_0 + 1)p_{n_0+1}(t) = \lambda n_0 p_{n_0}(t) = \lambda n_0 \mathrm{e}^{-\lambda n_0 t},$$

which is a standard first-order equation of the integrating-factor type. This can be solved to give

$$p_{n_0+1}(t) = \mathrm{e}^{-\lambda n_0 t}(1 - \mathrm{e}^{-\lambda t}).$$

However, it is easier to use the probability generating function method, which is also more general in its applicability. This generating function was first introduced in Section 1.9, and used also in Section 5.5 for the Poisson process.

Let

$$G(s, t) = \sum_{n=0}^{\infty} p_n(t)s^n, \tag{6.3}$$

but do not include the initial condition at this stage: in other words, do not put $p_n(t) = 0$ for $n < n_0$. The set of equations in (6.1) must be expressed in terms of the probability generating function $G(s, t)$ and its derivatives. Multiply both sides of (6.1) by s^n and sum from the lowest value of n to infinity as appropriate. Then

$$\sum_{n=0}^{\infty} \frac{\mathrm{d}p_n(t)}{\mathrm{d}t}s^n = \lambda \sum_{n=2}^{\infty}(n-1)p_{n-1}(t)s^n - \lambda \sum_{n=1}^{\infty} n p_n(t)s^n. \tag{6.4}$$

Now

$$\frac{\partial G(s, t)}{\partial t} = \sum_{n=0}^{\infty} \frac{\mathrm{d}p_n(t)}{\mathrm{d}t}s^n, \tag{6.5}$$

and

$$\frac{\partial G(s, t)}{\partial s} = \sum_{n=0}^{\infty} n p_n(t)s^{n-1}. \tag{6.6}$$

Thus, on the right-hand side of (6.4),

$$\sum_{n=2}^{\infty}(n-1)p_{n-1}(t)s^n = \sum_{m=1}^{\infty}mp_m(t)s^{m+1} = s^2\frac{\partial G(s,t)}{\partial s},$$

and

$$\sum_{n=1}^{\infty}np_n(t)s^n = s\frac{\partial G(s,t)}{\partial s}.$$

Equation (6.4) can be replaced therefore by the *partial differential equation*

$$\frac{\partial G(s,t)}{\partial t} = \lambda s^2\frac{\partial G(s,t)}{\partial s} - \lambda s\frac{\partial G(s,t)}{\partial s} = \lambda s(s-1)\frac{\partial G(s,t)}{\partial s}. \qquad (6.7)$$

We now have to solve this generating function equation. The initial condition $p_{n_0}(0) = 1$ translates into

$$G(s,0) = s^{n_0} \qquad (6.8)$$

for the generating function.

6.3 Birth process: generating function equation

We shall look at the solution of equations (6.7) and (6.8) in some detail since the method is used in a number of other applications. A change of variable is applied to (6.7) to remove the term $\lambda s(s-1)$. We achieve this by putting

$$\frac{ds}{dz} = \lambda s(s-1), \qquad (6.9)$$

and we shall assume that $0 < s < 1$ (for convergence reasons we are interested in the series for $G(s,t)$ for small s). This is a *first-order separable equation*, which can be separated and integrated as follows:

$$\int\frac{ds}{s(1-s)} = -\lambda dz = -\lambda z$$

(the value of any constant of integration is immaterial so we set it to a convenient value, in this case, to zero). Partial fractions are required on the left-hand side. Thus

$$\int\left[\frac{1}{s} + \frac{1}{1-s}\right]ds = -\lambda z,$$

or

$$\ln\left[\frac{s}{1-s}\right] = -\lambda z.$$

The solution of this equation for s gives

$$\frac{s}{1-s} = e^{-\lambda z} \qquad \text{or} \qquad s = \frac{1}{1+e^{\lambda z}}. \qquad (6.10)$$

Let

$$Q(z,t) = G(s,t) = G(1/(1+e^{\lambda z}), t).$$

We can check that

$$\frac{\partial Q(z,t)}{\partial z} = \frac{\partial G(1/(1+e^{\lambda z}),t)}{\partial z} = \frac{\partial G(s,t)}{\partial s}\cdot\frac{ds}{dz} = \lambda s(s-1)\frac{\partial G(s,t)}{\partial s},$$

using the chain rule in differentiation. Thus, equation (6.7) becomes

$$\frac{\partial Q(z,t)}{\partial t} = \frac{\partial Q(z,t)}{\partial z}.$$

The general solution of this equation for $Q(z,t)$ is any differentiable function w of $z+t$, that is, $Q(z,t) = w(z+t)$. This can be verified since

$$\frac{\partial}{\partial z}Q(z,t) = \frac{\partial}{\partial z}w(z+t) = w'(z+t).\frac{\partial(z+t)}{\partial z} = w'(z+t),$$

$$\frac{\partial}{\partial t}Q(z,t) = \frac{\partial}{\partial t}w(z+t) = w'(z+t).\frac{\partial(z+t)}{\partial t} = w'(z+t),$$

where $w'(z+t)$ means $dw(z+t)/d(z+t)$, in other words the derivative of w with respect to its argument. Note that if w is a function of $z+t$, then

$$w'(z+t) = \frac{\partial w(z+t)}{\partial z} = \frac{\partial w(z+t)}{\partial t}.$$

The function w is determined by the initial conditions. Thus, from (6.8) and (6.10),

$$G(s,0) = s^{n_0} = \frac{1}{(1+e^{\lambda z})^{n_0}} = w(z) = Q(z,0). \tag{6.11}$$

The function is defined in the middle of (6.11). Thus, by changing the argument in w from z to $z+t$, we obtain

$$G(s,t) = Q(z,t) = w(z+t) = \frac{1}{[1+e^{\lambda(z+t)}]^{n_0}}$$

$$= \frac{1}{[1+\frac{(1-s)}{s}e^{\lambda t}]^{n_0}} = \frac{s^{n_0}e^{-\lambda n_0 t}}{[1-(1-e^{-\lambda t})s]^{n_0}}, \tag{6.12}$$

using the change of variable (6.10) again. The probability generating function for the simple birth process is given by (6.12). The individual probabilities are the coefficients of s^n in the power series expansion of (6.12), which can be obtained by applying the binomial theorem to the denominator to derive a power series in s. Thus,

$$G(s,t) = \frac{s^{n_0}e^{-\lambda n_0 t}}{[1-(1-e^{-\lambda t})s]^{n_0}}$$

$$= s^{n_0}e^{-\lambda n_0 t}\left[1 + \frac{n_0}{1!}(1-e^{-\lambda t})s + \frac{n_0(n_0+1)}{2!}(1-e^{-\lambda t})^2 s^2 + \cdots\right]$$

$$= s^{n_0} e^{-\lambda n_0 t} \sum_{m=0}^{\infty} \binom{m + n_0}{n_0 - 1} (1 - e^{-\lambda t})^m s^m,$$

$$= e^{-\lambda n_0 t} \sum_{n=n_0}^{\infty} \binom{n - 1}{n_0 - 1} (1 - e^{-\lambda t})^{n - n_0} s^{n - n_0}, \tag{6.13}$$

where

$$\binom{r}{s} = \frac{r!}{s!(r - s)!}$$

is the *binomial coefficient*. If $n_0 = 1$, then

$$\binom{n - 1}{0} = 1,$$

for all n. Finally, from (6.13), the coefficients of the powers of s imply that, since the leading power is s^{n_0},

$$p_n(t) = 0, \qquad (n < n_0),$$

$$p_n(t) = \binom{n - 1}{n_0 - 1} e^{-\lambda n_0 t} (1 - e^{-\lambda t})^{n - n_0}, \qquad (n \geq n_0),$$

which is a Pascal distribution with parameters $(n_0, e^{-\lambda t})$ (see Section 1.7). Some graphs of the first four probabilities are shown in Figure 6.1 for the birth process starting with just one individual.

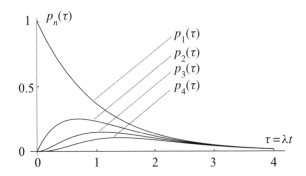

Fig. 6.1 *The probabilities $p_n(\tau)$ shown for an initial population $n_0 = 1$ and $n = 1, 2, 3, 4$: a dimensionless time scale $\tau = \lambda t$ has been used.*

The mean population size at time t is given by (equation (5.18))

$$\mu(t) = G_s(1, t) = \frac{\partial}{\partial s} \left[\frac{s^{n_0} e^{-\lambda n_0 t}}{[1 - (1 - e^{-\lambda t})s]^{n_0}} \right]_{s=1}$$

$$= \left[\frac{n_0 s^{n_0 - 1} e^{-\lambda n_0 t}}{[1 - (1 - e^{-\lambda t})s]^{n_0}} + \frac{n_0 s^{n_0} e^{-\lambda n_0 t} (1 - e^{-\lambda t})}{[1 - (1 - e^{-\lambda t})s]^{n_0 + 1}} \right]_{s=1}$$

$$= n_0 (1 - e^{-\lambda t}) e^{\lambda t} + n_0 = n_0 e^{\lambda t},$$

or can be deduced directly from the Pascal mean. The expected population size grows exponentially with time.

6.4 The death process

In this case, there are no births and the population numbers decline through deaths. Again we assume that the probability that any individual dies in a short time interval δt is $\mu \delta t$, where μ is the death rate. The probability that a death occurs in a population of size n is $\mu n \delta t$ and, as before, the probability of multiple deaths is assumed to be negligible. By arguments similar to those for the birth process

$$p_0(t + \delta t) = [\mu \delta t + o(\delta t)] p_1(t),$$

$$p_n(t + \delta t) = [\mu(n+1)\delta t + o(\delta t)] p_{n+1}(t) + [1 - \mu n \delta t - o(\delta t)] p_n(t).$$

Thus

$$\frac{p_0(t + \delta t) - p_0(t)}{\delta t} = \mu p_1(t) + o(1),$$

$$\frac{p_n(t + \delta t) - p_n(t)}{\delta t} = \mu(n+1) p_{n+1}(t) - \mu n p_n(t) + o(1), \qquad (n \geq 1),$$

which become, in the limit $\delta t \to 0$,

$$\left. \begin{aligned} \frac{\mathrm{d}p_0(t)}{\mathrm{d}t} &= \mu p_1(t), \\ \frac{\mathrm{d}p_n(t)}{\mathrm{d}t} &= \mu(n+1) p_{n+1}(t) - \mu n p_n(t). \end{aligned} \right\} \tag{6.14}$$

If the initial population size is n_0 at time $t = 0$, then $p_{n_0}(0) = 1$ and also $p_n(t) = 0$ for $n > n_0$ since this is a *death* process. Consequently, we expect the probability generating function to be a finite series. Multiply each equation in (6.14) by s^n and sum over $n \geq 0$:

$$\sum_{n=0}^{\infty} \frac{\mathrm{d}p_n(t)}{\mathrm{d}t} s^n = \mu \sum_{n=0}^{\infty} (n+1) p_{n+1}(t) s^n - \mu \sum_{n=1}^{\infty} n p_n(t) s^n. \tag{6.15}$$

By (6.6)

$$\sum_{n=0}^{\infty} (n+1) p_{n+1}(t) s^n = \sum_{n=1}^{\infty} n p_n(t) s^{n-1} = \frac{\partial G(s,t)}{\partial s}.$$

Hence, using (6.5) and (6.6) again, equation (6.15) becomes

$$\frac{\partial G(s,t)}{\partial t} = \mu \frac{\partial G(s,t)}{\partial s} - \mu s \frac{\partial G(s,t)}{\partial s} = \mu(1-s) \frac{\partial G(s,t)}{\partial s}. \tag{6.16}$$

The method of solution now tracks that given for the birth process in the previous section. The difference between (6.7) and (6.16) lies solely in the different change of variable

$$\frac{\mathrm{d}s}{\mathrm{d}z} = \mu(1 - s).$$

On this occasion

$$\int \frac{ds}{1-s} = \int \mu dz = \mu z.$$

Hence, for $0 < s < 1$,

$$-\ln(1-s) = \mu z, \quad \text{or} \quad s = 1 - e^{-\mu z}.$$

For the death process we let

$$Q(z, t) = G(s, t) = G(1 - e^{-\mu z}, t),$$

where $Q(z, t)$ now satisfies

$$\frac{\partial Q(z, t)}{\partial t} = \frac{\partial Q(z, t)}{\partial z}$$

which is the same partial differential equation as for the birth process but with a *different* change of variable. As before, the general solution is

$$G(s, t) = w(t + z).$$

If the initial population size is n_0, then $G(s, 0) = s^{n_0}$. Hence

$$G(s, 0) = s^{n_0} = (1 - e^{-\mu z})^{n_0} = w(z) = Q(z, 0).$$

Hence

$$\begin{aligned}
G(s, t) = Q(z, t) = w(z + t) &= (1 - e^{-\mu(z+t)})^{n_0} \\
&= [1 - e^{-\mu t}(1 - s)]^{n_0} \\
&= (1 - e^{-\mu t})^{n_0} \left[1 + \frac{se^{-\mu t}}{1 - e^{-\mu t}} \right]^{n_0}.
\end{aligned} \qquad (6.17)$$

The individual probabilities can be found by expanding the right-hand side of this formula for $G(s, t)$ using the binomial theorem (see Problem 6.3).

Example 6.1. Find the mean $\mu(t)$ of the population at time t. Show that $\mu(t)$ satisfies the differential equation

$$\frac{d\mu(t)}{dt} = -\mu\mu(t),$$

where $p_{n_0}(0) = 1$. Interpret the result.

The mean is given by

$$\begin{aligned}
\mu(t) = G_s(1, t) &= \frac{\partial}{\partial s} \left[1 - e^{-\mu t}(1 - s) \right]^{n_0} \Big|_{s=1} \\
&= n_0 e^{-\mu t} [1 - e^{-\mu t}(1 - s)]^{n_0 - 1} \Big|_{s=1} \\
&= n_0 e^{-\mu t}.
\end{aligned}$$

Hence

$$\frac{\mathrm{d}\boldsymbol{\mu}(t)}{\mathrm{d}t} = -\mu n_0 \mathrm{e}^{-\mu t} = -\mu\boldsymbol{\mu}(t)$$

with $\boldsymbol{\mu}(0) = n_0$.

Suppose that we consider a *deterministic model* of a population in which the population size $n(t)$ is a continuous function of time rather than a random variable with *discrete* values. This is a justifiable approximation at least for large populations. We could then model the population change by postulating that the rate of decrease of the population is proportional to the current population size $n(t)$. Thus

$$\frac{\mathrm{d}n(t)}{\mathrm{d}t} \propto n(t) \qquad \text{or} \qquad \frac{\mathrm{d}n(t)}{\mathrm{d}t} = -\mu n(t)$$

where μ is the death-rate. This is the *Malthus model* for the death process. We can deduce from this that the mean of the stochastic process satisfies the differential equation of the simple deterministic model. This provides some justification for using deterministic models in large populations. □

It is easy to calculate from the probability generating function the probability of *extinction* at time t. It is the probability that the population size is zero at time t, namely,

$$p_0(t) = G(0, t) = [1 - \mathrm{e}^{-\mu t}]^{n_0}.$$

The probability of *ultimate extinction* is

$$\lim_{t\to\infty} p_0(t) = \lim_{t\to\infty} G(0, t) = \lim_{t\to\infty} [1 - \mathrm{e}^{-\mu t}]^{n_0} = 1,$$

since $\lim_{t\to\infty} \mathrm{e}^{-\mu t} = 0$. In other words, the probability of ultimate extinction is certain, as we would expect in a death process.

6.5 The combined birth and death process

Both processes are now combined into one with a birth-rate λ and a death-rate μ. We can construct the differential-difference equations from (6.1) and (6.14). Thus, $\mathrm{d}p_n(t)/\mathrm{d}t$ is simply the sum of the right-hand sides of these equations. The result is

$$\left.\begin{array}{c} \dfrac{\mathrm{d}p_0(t)}{\mathrm{d}t} = \mu p_1(t) \\[2mm] \dfrac{\mathrm{d}p_n(t)}{\mathrm{d}t} = \lambda(n-1)p_{n-1}(t) - (\lambda+\mu)np_n(t) + \mu(n+1)p_{n+1}(t), \ (n \geq 1) \end{array}\right\}$$

$$(6.18)$$

Similarly, the partial differential equation of the probability generating function can be reconstructed from (6.7) and (6.16):

$$\frac{\partial G(s, t)}{\partial t} = \lambda s(s - 1)\frac{\partial G(s, t)}{\partial s} + \mu(1 - s)\frac{\partial G(s, t)}{\partial s}$$

$$= (\lambda s - \mu)(s - 1)\frac{\partial G(s, t)}{\partial s}. \qquad (6.19)$$

There are two cases to consider, namely $\lambda \neq \mu$ and $\lambda = \mu$.

(a) $\lambda \neq \mu$

The change of variable is

$$\frac{ds}{dz} = (\lambda s - \mu)(s - 1).$$

Hence

$$
\begin{aligned}
z = \int dz &= \int \frac{ds}{(\lambda s - \mu)(s - 1)} \\
&= \frac{1}{\lambda} \int \frac{ds}{(\frac{\mu}{\lambda} - s)(1 - s)} \\
&= \frac{1}{\lambda - \mu} \int \left[\frac{1}{\frac{\mu}{\lambda} - s} - \frac{1}{1 - s} \right] ds, \qquad \text{(using partial fractions)} \\
&= \frac{1}{\lambda - \mu} \ln \left[\frac{1 - s}{\frac{\mu}{\lambda} - s} \right], \qquad (0 < s < \frac{\mu}{\lambda} \text{ and } 0 < s < 1). \qquad (6.20)
\end{aligned}
$$

The inversion of this formula gives

$$s = \frac{\lambda - \mu e^{(\lambda - \mu)z}}{\lambda - \lambda e^{(\lambda - \mu)z}}. \qquad (6.21)$$

If the initial population size is n_0, then

$$G(s, 0) = s^{n_0} = \left[\frac{\lambda - \mu e^{(\lambda - \mu)z}}{\lambda - \lambda e^{(\lambda - \mu)z}} \right]^{n_0} = w(z) = Q(z, 0).$$

Hence

$$G(s, t) = Q(z, t) = w(z + t) = \left[\frac{\lambda - \mu e^{(\lambda - \mu)(z+t)}}{\lambda - \lambda e^{(\lambda - \mu)(z+t)}} \right]^{n_0}. \qquad (6.22)$$

From (6.20)

$$e^{(\lambda - \mu)z} = \frac{1 - s}{\frac{\mu}{\lambda} - s} = \frac{\lambda(1 - s)}{\mu - \lambda s}.$$

Finally, elimination of z in (6.22) leads to

$$G(s, t) = \left[\frac{\mu(1 - s) - (\mu - \lambda s)e^{-(\lambda - \mu)t}}{\lambda(1 - s) - (\mu - \lambda s)e^{-(\lambda - \mu)t}} \right]^{n_0}. \qquad (6.23)$$

The expected population size is, at time t, for $\lambda \neq \mu$,

$$\mu(t) = \sum_{n=1}^{\infty} n p_n(t) = G_s(1, t)$$

$$= \frac{n_0(-\mu + \lambda e^{-(\lambda-\mu)t})}{-(\mu - \lambda)e^{-(\lambda-\mu)t}} - \frac{n_0(-\lambda + \lambda e^{(\lambda-\mu)t})}{-(\mu - \lambda)e^{-(\lambda-\mu)t}}$$

$$= n_0 e^{(\lambda-\mu)t},$$

using (6.23).

(b) $\lambda = \mu$

In this case (6.19) becomes

$$\frac{\partial G(s, t)}{\partial t} = \lambda(1 - s)^2 \frac{\partial G(s, t)}{\partial s}.$$

Let

$$\frac{ds}{dz} = \lambda(1 - s)^2.$$

Then the change of variable is

$$z = \frac{1}{\lambda} \int \frac{ds}{(1 - s)^2} = \frac{1}{\lambda(1 - s)} \quad \text{or} \quad s = \frac{\lambda z - 1}{\lambda z}.$$

It follows that

$$f(z) = s^{n_0} = \left[\frac{\lambda z - 1}{\lambda z} \right]^{n_0}.$$

Finally, the probability generating function for the special case in which the birth and death rates are the same is given by

$$G(s, t) = \left[\frac{\lambda(z + t) - 1}{\lambda(z + t)} \right]^{n_0} = \left[\frac{1 + (\lambda t - 1)(1 - s)}{1 + \lambda t(1 - s)} \right]^{n_0}. \tag{6.24}$$

The expected value of the population size at time t in the case $\lambda = \mu$ is left as a problem at the end of the chapter. The probability of extinction at time t for the case $\lambda = \mu$ is, from (6.24),

$$p_0(t) = G(0, t) = \left[\frac{1 + (\lambda t - 1)}{1 + \lambda t} \right]^{n_0} = \left[\frac{\lambda t}{1 + \lambda t} \right]^{n_0}.$$

As t becomes large this value approaches 1 since

$$\lim_{t \to \infty} p_0(t) = \lim_{t \to \infty} \left[\frac{1}{1 + \frac{1}{\lambda t}} \right]^{n_0} = 1.$$

We obtain the interesting conclusion that, if the birth and death rates are in balance, then ultimate extinction is certain.

That the probability generating functions in equations (6.23) and (6.24) are powers of certain functions of s and t is not surprising if we recollect the definition of the generating function (see Section 1.9) as

$$G(s, t) = \mathbf{E}[s^{N(t)}],$$

where $N(t)$ is a random variable of the number of individuals in the population at time t. Suppose that we identify the n_0 individuals of the initial population and let $N_i(t)$ represent the number of descendants of the ith individual so that

$$N(t) = N_1(t) + N_2(t) + \cdots + N_{n_0}(t).$$

Hence

$$G(s, t) = \mathbf{E}[s^{N(t)}] = \mathbf{E}[s^{N_1(t)+N_2(t)+\cdots+N_{n_0}(t)}].$$

Since the $N_i(t)$ are independent and identically distributed (iid), it follows that

$$G(s, t) = \mathbf{E}[s^{N_1(t)}]\mathbf{E}[s^{N_2(t)}]\ldots\mathbf{E}[s^{N_{n_0}(t)}] = [\mathbf{E}[s^{N_i(t)}]]^{n_0}$$

for any i (see Section 1.9(e)). Hence $G(s, t)$ must be the n_0th power of some function of s and t.

Example 6.2. In the birth and death process with $\lambda = \mu$, show that the mean time to extinction is infinite for any initial population size n_0.

We require the probability distribution $F(t)$ for the time T_{n_0} to extinction, that is,

$$F(t) = \mathbf{P}\{T_{n_0} \le t\}$$
$$= p_0(t) = \left[\frac{\lambda t}{1 + \lambda t}\right]^{n_0}$$

from (6.25). The density function $f(t)$ of this distribution is

$$f(t) = \frac{dF(t)}{dt} = \frac{n_0\lambda^{n_0}t^{n_0-1}}{(1 + \lambda t)^{n_0}} - \frac{n_0(\lambda t)^{n_0}\lambda}{(1 + \lambda t)^{n_0+1}} = \frac{n_0\lambda^{n_0}t^{n_0-1}}{(1 + \lambda t)^{n_0+1}}.$$

The expected value of the random variable T_{n_0} will be given by

$$\mathbf{E}(T_{n_0}) = \int_0^\infty tf(t)dt$$

only if the integral on the right *converges*. In this case

$$tf(t) = \frac{n_0\lambda^{n_0}t^{n_0}}{(1 + \lambda t)^{n_0+1}}.$$

For *large t*, $1 + \lambda t \approx \lambda t$ and

$$tf(t) \approx \frac{n_0 \lambda^{n_0} t^{n_0}}{(\lambda t)^{n_0+1}} = \frac{n_0}{\lambda} \frac{1}{t}.$$

Although $tf(t) \to 0$ as $t \to \infty$, it is too slow for the integral to converge. For example, the integral

$$\int_1^\tau \frac{dt}{t} = \left[\ln t \right]_1^\tau = \ln \tau \to \infty$$

as $\tau \to \infty$, has the same behaviour for large t and this integral *diverges*. We conclude that $E(T_{n_0}) = \infty$.

This is another interesting result to add to that of the probability of extinction. Notwithstanding that extinction is ultimately certain, we expect it to take an infinite time. A graph of $p_0(t)$ is shown in Figure 6.2.

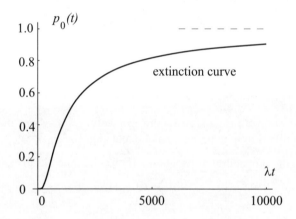

Fig. 6.2 *The graph shows the behaviour of the probability of extinction* $p_0(t) = [\lambda t/(1 + \lambda t)]^{n_0}$ *against* λt *for an initial population size of* $n_0 = 1000$ *in the case* $\lambda = \mu$. *The actual time scale will depend on the parameter* λ.

$\square$

6.6 General population processes

The previous models for birth and death processes assume that the total rates of births and deaths are proportional to the population size. In more general population processes, the rates can be more general functions of n which perhaps could represent more realistic models such as higher death rates for overcrowded larger populations.

Assume that the birth and death rates are λ_n and μ_n respectively. Using arguments that lead to (6.2) and (6.14) but with the new coefficients, the governing differential-difference equations are replaced by

$$\left.\begin{array}{l} \dfrac{dp_0(t)}{dt} = -\lambda_0 p_0(t) + \mu_1 p_1(t) \\[2mm] \dfrac{dp_n(t)}{dt} = \lambda_{n-1} p_{n-1}(t) - (\lambda_n + \mu_n)p_n(t) + \mu_{n+1}p_{n+1}(t), \quad (n \geq 1) \end{array}\right\},$$

(6.25)

but note that λ_0 must be included since we cannot assume it is zero. The simple birth and death process is given by the particular case in which $\lambda_n = n\lambda$ and $\mu_n = n\mu$ are *linear* functions of n. The Poisson process of the previous chapter corresponds to $\lambda_n = \lambda$ and $\mu_n = 0$.

The following example develops a population model in which the birth-rate is constant, perhaps sustained by immigration, and the death rate is simple and given by $\mu_n = n\mu$. In other words, there are no indigenous births.

Example 6.3. In a population model, the immigration rate $\lambda_n = \lambda$, a constant, and the death-rate is $\mu_n = n\mu$. Assuming that the initial population size is n_0, find the probability generating function for this process, and find the mean population size at time t.

Insert λ_n and μ_n into (6.25), multiply by s^n and sum over $n = 0, 1, 2, \ldots$. The result is

$$\sum_{n=0}^{\infty} \frac{dp_n(t)}{dt} s^n = \lambda \sum_{n=1}^{\infty} p_{n-1}(t)s^n - \lambda \sum_{n=0}^{\infty} p_n(t)s^n - \mu \sum_{n=1}^{\infty} n p_n(t)s^n$$

$$+ \mu \sum_{n=0}^{\infty} (n+1)p_{n+1}(t)s^n. \tag{6.26}$$

Using the series for $G(s, t)$, and those given by (6.5) and (6.6) for its partial derivatives, equation (6.26) becomes

$$\frac{\partial G(s, t)}{\partial t} = \lambda s G(s, t) - \lambda G(s, t) - \mu s \frac{\partial G(s, t)}{\partial s} + \mu \frac{\partial G(s, t)}{\partial s}$$

$$= \lambda(s-1)G(s, t) + \mu(1-s)\frac{\partial G(s, t)}{\partial s}. \tag{6.27}$$

This is a partial differential equation of a type we have not encountered previously since it includes the term $G(s, t)$ in addition to the partial derivatives. This term can be removed by introducing a new function, $S(s, t)$, through the transformation

$$G(s, t) = e^{\lambda s/\mu} S(s, t).$$

Then

$$\frac{\partial G(s, t)}{\partial t} = e^{\lambda s/\mu} \frac{\partial S(s, t)}{\partial t}, \qquad \frac{\partial G(s, t)}{\partial s} = e^{\lambda s/\mu} \frac{\partial S(s, t)}{\partial s} + \frac{\lambda}{\mu} e^{\lambda s/\mu} S(s, t).$$

Substitution of the partial derivatives into (6.27) leads to

$$\frac{\partial S(s,t)}{\partial t} = \mu(1-s)\frac{\partial S(s,t)}{\partial s}$$

for $S(s,t)$. In fact, $S(s,t)$ now satisfies the same equation as that for the death process in Section 4.4. However, the initial condition is different since, for an initial population of n_0,

$$G(s,0) = s^{n_0}, \qquad \text{but} \qquad S(s,0) = e^{-\lambda s/\mu} s^{n_0}.$$

The change of variable is

$$s = 1 - e^{-\mu z}.$$

Hence

$$S(s,0) = e^{-\lambda s/\mu} s^{n_0} = e^{-\lambda(1-e^{-\mu z})/\mu}(1 - e^{-\mu z})^{n_0} = w(z),$$

from which it follows that

$$\begin{aligned}
G(s,t) &= e^{\lambda s/\mu} w(z+t) \\
&= e^{\lambda s/\mu} \exp[-\lambda(1 - e^{-\mu(z+t)})/\mu][1 - e^{-\mu(z+t)}]^{n_0} \\
&= e^{\lambda s/\mu} \exp[-\lambda(1 - (1-s)e^{-\mu t})/\mu][1 - (1-s)e^{-\mu t}]^{n_0}.
\end{aligned}$$

Notice that, unlike the probability generating function for the birth and death process in Section 6.5, this generating function is *not* an n_0 power of a function of s and t. The reason for this difference lies in the immigration, or, equivalently, the birth-rate, which, for any individual depends on n, the number of individuals in the population at any time. This means that the random variables $N_i(t)$, $(i = 1, 2, \ldots n_0)$, are not independent, which is a requirement of the result in Section 1.9(e).

To find the mean, take the logarithm of both sides of this equation:

$$\ln G(s,t) = \frac{\lambda s}{\mu} - \frac{\lambda}{\mu}[1 - (1-s)e^{-\mu t}] + n_0 \ln[1 - (1-s)e^{-\mu t}].$$

Differentiate with respect to s and put $s = 1$:

$$\frac{G_s(1,t)}{G(1,t)} = \frac{\lambda}{\mu} - \frac{\lambda}{\mu} e^{-\mu t} + n_0 e^{-\mu t}.$$

Now $G(1,t) = 1$ and the mean is therefore

$$\mu(t) = G_s(1,t) = \frac{\lambda}{\mu}(1 - e^{-\mu t}) + n_0 e^{-\mu t}.$$

As $t \to \infty$, the long-term mean approaches the ratio of the rates λ/μ. □

The birth and death rates can be functions of time in addition to population size. This might reflect declining fertility or seasonal variations in births or deaths. The setting up of the equation for the probability generating function presents no particular problem, although its solution inevitably becomes more complicated. The following example illustrates a birth process with a declining birth-rate.

Example 6.4. A colony of bacteria grows without deaths with a birth-rate that is time-dependent. The rate declines exponentially with time according to $\lambda(t) = \alpha e^{-\beta t}$, where α and β are constants. Find the mean population size at time t given that the initial size of the colony is n_0.

For a time-varying birth-rate the construction of the equation for the probability generating function is still given by (6.7) but with λ as a function of time. Thus, $G(s, t)$ satisfies

$$\frac{\partial G(s, t)}{\partial t} = \lambda(t)s(s - 1)\frac{\partial G(s, t)}{\partial s} = \alpha e^{-\beta t}s(s - 1)\frac{\partial G(s, t)}{\partial s}$$

with $G(s, 0) = s^{n_0}$. We now apply the *double* change of variable

$$\frac{ds}{dt} = \alpha s(s - 1), \qquad \frac{dt}{d\tau} = e^{\beta t}.$$

As in Section 6.3,

$$s = \frac{1}{1 + e^{\alpha z}},$$

whilst

$$\tau = \frac{1}{\beta}(1 - e^{-\beta t})$$

so that $\tau = 0$ when $t = 0$. With these changes of variable

$$Q(z, \tau) = G(s, t) = G(1/(1 + e^{\alpha z}), -\beta^{-1}\ln(1 - \beta\tau)),$$

where $Q(z, \tau)$ satisfies

$$\frac{\partial Q(z, \tau)}{\partial \tau} = \frac{\partial Q(z, \tau)}{\partial z}.$$

As in Section 6.3

$$Q(z, \tau) = w(z + \tau)$$

for any differentiable function w. The initial condition determines w through

$$G(s, 0) = s^{n_0} = \frac{1}{(1 + e^{\alpha z})^{n_0}} = w(z) = Q(z, 0).$$

Hence

$$G(s, t) = Q(z, \tau) = w(z + \tau) = [1 + e^{\alpha(z+\tau)}]^{-n_0}$$

$$= [1 + \left(\frac{1 - s}{s}\right)e^{\alpha(1 - e^{-\beta t})/\beta}]^{-n_0}$$

$$= s^{n_0}[1 + (1 - s)e^{\alpha(1 - e^{-\beta t})/\beta}]^{-n_0}.$$

The mean population size at time t is

$$\mu(t) = G_s(1, t) = \frac{\partial}{\partial s} \left[\frac{s^{n_0}}{[1 + (1 - s)e^{\alpha(1-e^{-\beta t})/\beta}]^{n_0}} \right]_{s=1} = n_0 e^{\alpha(1-e^{-\beta t})/\beta}.$$

As $t \to \infty$, then the mean population approaches the limit $n_0 e^{\alpha/\beta}$. □

Problems

6.1. A colony of cells grows from a single cell. The probability that a cell divides in a time interval δt is

$$\lambda \delta t + o(\delta t).$$

There are no deaths. Show that the probability generating function for this death process is

$$G(s, t) = \frac{se^{-\lambda t}}{1 - (1 - e^{-\lambda t})s}.$$

Find the probability that the original cell has not divided at time t, the mean population size at time t and its variance (see Problem 5.4 for the variance formula using the probability generating function).

6.2. A simple birth process has a constant birth-rate λ. Show that its mean population size $\mu(t)$ satisfies the differential equation

$$\frac{d\mu(t)}{dt} = \lambda \mu(t).$$

How can this result be interpreted in terms of a deterministic model for a birth process?

6.3. The probability generating function for a simple death process with death-rate μ and initial population size n_0 is given by

$$G(s, t) = (1 - e^{-\mu t})^{n_0} \left[1 + \frac{se^{-\mu t}}{1 - e^{-\mu t}} \right]^{n_0}$$

(see Equation (6.17)). Using the binomial theorem find the probability $p_n(t)$ for $n \le n_0$. If n_0 is an *even* number, find the probability that the population has halved by time t. A large number of experiments were undertaken with live samples with a variety of initial populations drawn from a common source and the times of the halving of deaths were recorded for each sample. What would be the most likely time for the population to halve?

6.4. A birth process has a probability generating function $G(s, t)$ given by

$$G(s, t) = \frac{s}{e^{\lambda t} + s(1 - e^{-\lambda t})}.$$

    (a) What is the initial population size?

    (b) Find the probability that the population size is n at time t.

    (c) Find the mean and variance of the population at time t.

6.5. A random process has the probability generating function

$$G(s,t) = \left(\frac{2+st}{2+t}\right)^r,$$

where r is a positive integer. What is the initial state of the process? Find the probability $p_n(t)$ associated with the generating function. What is $p_r(t)$? Show that the mean associated with $G(s,t)$ is

$$\mu(t) = \frac{rt}{2+t}.$$

6.6. In a simple birth and death process with unequal birth and death-rates λ and μ, the probability generating function is given by

$$G(s,t) = \left[\frac{\mu(1-s) - (\mu - \lambda s)e^{-(\lambda-\mu)t}}{\lambda(1-s) - (\mu - \lambda s)e^{-(\lambda-\mu)t}}\right]^{n_0},$$

for an initial population size n_0 (see Equation (6.23)).

    (a) Find the mean population at time t.

    (b) Find the probability of extinction at time t.

    (c) Show that, if $\lambda < \mu$, then the probability of ultimate extinction is 1. What is it if $\lambda > \mu$?

    (d) Find the variance of the population size.

6.7. In a population model, the immigration rate $\lambda_n = \lambda$, a constant, and the death rate $\mu_n = n\mu$. For an initial population size n_0, the probability generating function is (Example 6.3)

$$G(s,t) = e^{\lambda s/\mu}\exp[-\lambda(1-(1-s)e^{-\mu t})/\mu][1-(1-s)e^{-\mu t}]^{n_0}.$$

Find the probability that extinction occurs at time t. What is the probability of ultimate extinction?

6.8. In a general birth and death process, a population is maintained by immigration at a constant rate λ, and the death-rate is $n\mu$. Using the differential-difference equations (6.26) directly, obtain the differential equation

$$\frac{d\mu(t)}{dt} + \mu\mu(t) = \lambda,$$

for the mean population size $\mu(t)$. Solve this equation assuming an initial population n_0 and compare the answer with that given in Example 6.3.

6.9. In a death process, the probability of a death when the population size is $n \neq 0$ is a constant μ but obviously zero if the population is zero. Verify that, if the initial population is n_0, then $p_n(t)$, the probability that the population size is n at time t is given by

$$p_n(t) = \frac{(\mu t)^{n_0 - n}}{(n_0 - n)!} e^{-\mu t}, \qquad (1 \leq n \leq n_0),$$

$$p_0(t) = \frac{\mu^{n_0}}{(n_0 - 1)!} \int_0^t s^{n_0 - 1} e^{-\mu s} \, ds.$$

Show that the mean time to extinction is n_0 / μ.

6.10. In a birth and death process, the birth and death rates are given by

$$\lambda_n = n\lambda + \alpha, \qquad \mu_n = n\mu,$$

where α represents a constant immigration rate. Show that the probability generating function $G(s, t)$ of the process satisfies

$$\frac{\partial G(s, t)}{\partial t} = (\lambda s - \mu)(s - 1) \frac{\partial G(s, t)}{\partial s} + \alpha(s - 1)G(s, t).$$

Show also that, if

$$G(s, t) = (\mu - \lambda s)^{-\alpha/\lambda} S(s, t),$$

then $S(s, t)$ satisfies

$$\frac{\partial S(s, t)}{\partial t} = (\lambda s - \mu)(s - 1) \frac{\partial S(s, t)}{\partial s}.$$

Let the initial population size be n_0. Solve the partial differential equation for $S(s, t)$ using the method of Section 6.5 and confirm that

$$G(s, t) = \frac{(\mu - \lambda)^{\alpha/\lambda} [(\mu - \lambda s) - \mu(1 - s)e^{(\lambda - \mu)t}]^{n_0}}{[(\mu - \lambda s) - \lambda(1 - s)e^{(\lambda - \mu)t}]^{n_0 + (\alpha/\lambda)}}.$$

(Remember the modified initial condition for $S(s, t)$.)

Find $p_0(t)$, the probability that the population is zero at time t (since immigration takes place even when the population is zero there is no question of extinction in this process). Hence show that

$$\lim_{t \to \infty} p_0(t) = \left(\frac{\mu - \lambda}{\mu} \right)^{\alpha/\lambda}$$

if $\lambda < \mu$. What is the limit if $\lambda > \mu$?

The long-term behaviour of the process for $\lambda < \mu$ can be investigated by looking at the limit of the probability generating function as $t \to \infty$. Show that

$$\lim_{t \to \infty} G(s, t) = \left(\frac{\mu - \lambda}{\mu - \lambda s} \right)^{\alpha/\lambda}.$$

This is the probability generating function of a *stationary distribution* and it indicates that a balance has been achieved between the birth and immigration rates, and the death rate. What is the long term mean population?

If you want a further lengthy exercise, investigate the probability generating function in the special case $\lambda = \mu$.

6.11. In a birth and death process with immigration, the birth and death rates are respectively

$$\lambda_n = n\lambda + \alpha, \qquad\qquad \mu_n = n\mu.$$

Show directly from the differential-difference equations for $p_n(t)$, that the mean population size $\mu(t)$ satisfies the differential equation

$$\frac{d\mu(t)}{dt} = (\lambda - \mu)\mu(t) + \alpha.$$

Deduce the result

$$\mu(t) \to \frac{\alpha}{\mu - \lambda}$$

as $t \to \infty$ if $\lambda < \mu$. Discuss the design of a deterministic immigration model based on this equation.

6.12. In a simple birth and death process with equal birth and death rates λ, the initial population size has a Poisson distribution with probabilities

$$p_n(0) = e^{-\alpha}\frac{\alpha^n}{n!},$$

with parameter α. It could be thought of as a process in which the initial distribution has arisen as the result of some previous process. Find the probability generating function for this process, and confirm that the probability of extinction at time t is $\exp[-\alpha/(1 + \lambda t)]$ and that the mean population size is α for all t.

6.13. A birth and death process takes place as follows. A single bacterium is allowed to grow and is assumed to behave as a simple birth process with birth-rate λ for a time t_1 without any deaths. No further growth then takes place. The colony of bacteria is then allowed to die with the assumption that it is a simple death process with death-rate μ for a time t_2. Show that the probability of extinction after the total time $t_1 + t_2$ is

$$\sum_{n=1}^{\infty} e^{\lambda t_1}(1 - e^{-\lambda t_1})^{n-1}(1 - e^{-\mu t_2})^n.$$

Using the formula for the sum of a geometric series, show that this probability can be simplified to

$$\frac{e^{\mu t_2} - 1}{e^{\lambda t_1} + e^{\mu t_2} - 1}.$$

6.14. As in the previous problem a single bacterium grows as a simple birth process with rate λ and no deaths for a time τ. The colony numbers then decline as a simple death process with rate μ. Show that the probability generating function for the death process is

$$\frac{(1 - e^{-\mu t}(1 - s))e^{-\lambda \tau}}{1 - (1 - e^{-\lambda \tau})(1 - e^{-\mu t}(1 - s))},$$

where t is measured from the time τ. Show that the mean population size during the death process is $e^{\lambda \tau - \mu t}$.

6.15. For a simple birth and death process, the probability generating function (equation (6.23)) is given by

$$G(s, t) = \left[\frac{\mu(1 - s) - (\mu - \lambda s)e^{-(\lambda - \mu)t}}{\lambda(1 - s) - (\mu - \lambda s)e^{-(\lambda - \mu)t}} \right]^{n_0}$$

for an initial population of n_0. Assume that $\lambda > \mu$. Using a two-term binomial expansion, show that, for large t,

$$G(s, t) \approx \left(\frac{\mu}{\lambda} \right)^{n_0} \left[1 - n_0 \left(\frac{\mu - \lambda s}{1 - s} \right) \left(\frac{\lambda - \mu}{\lambda \mu} \right) e^{-(\lambda - \mu)t} \right].$$

By expanding $(1 - s)^{-1}$ in powers of s obtain the probabilities $p_n(t)$ for large t.

6.16. (An alternative method of solution for the probability generating function.) The general solution of the first-order partial differential equation

$$A(x, y, z)\frac{\partial z}{\partial x} + B(x, y, z)\frac{\partial z}{\partial y} = C(x, y, z)$$

is

$$f(u, v) = 0$$

where f is an arbitrary function, and $u(x, y, z) = c_1$ and $v(x, y, z) = c_2$ are two independent solutions of

$$\frac{dx}{A(x, y, z)} = \frac{dy}{B(x, y, z)} = \frac{dz}{C(x, y, z)}.$$

This is known as *Cauchy's method*.

Apply the method to the partial differential equation for the probability generating function for the simple birth and death process, namely (equation (6.19))

$$\frac{\partial G(s, t)}{\partial t} = (\lambda s - \mu)(s - 1)\frac{\partial G(s, t)}{\partial s},$$

by solving

$$\frac{ds}{(\lambda s - \mu)(1 - s)} = \frac{dt}{1} = \frac{dG}{0}.$$

Show that

$$u(s, t, G) = G = c_1, \qquad \text{and} \qquad v(s, t, P) = e^{-(\lambda-\mu)t}\left(\frac{1-s}{\frac{\mu}{\lambda}-s}\right) = c_2.$$

are two independent solutions. The general solution can be written in the form

$$G(s, t) = H\left[e^{-(\lambda-\mu)t}\left(\frac{1-s}{\frac{\mu}{\lambda}-s}\right)\right].$$

Here, H is a function determined by the initial condition $P(s, 0) \doteq s^{n_0}$. Find H and recover formula (6.22) for the probability generating function.

6.17. Apply Cauchy's method outlined in Problem 6.16 to the immigration model in Example 6.3. In this application the probability generating function satisfies

$$\frac{\partial G(s, t)}{\partial t} = \lambda(s - 1)G(s, t) + \mu(1 - s)\frac{\partial G(s, t)}{\partial s}.$$

Solve the equation assuming an initial population of n_0.

6.18. In a population sustained by immigration at rate λ with a simple death process with rate μ (see Example 6.3), the probability $p_n(t)$ satisfies (equation (6.25))

$$\frac{dp_0(t)}{dt} = -\lambda p_0(t) + \mu p_1(t),$$

$$\frac{dp_n(t)}{dt} = \lambda p_{n-1}(t) - (\lambda + n\mu)p_n(t) + (n + 1)\mu p_{n+1}(t).$$

Investigate the steady-state behaviour of the system by assuming that $p_n(t) \to p_n$, $dp_n(t)/dt \to 0$ as $t \to \infty$. Show that the resulting difference equations for what is known as the corresponding *stationary process*

$$-\lambda p_0 + \mu p_1 = 0,$$

$$\lambda p_{n-1} - (\lambda + n\mu)p_n + (n + 1)\mu p_{n+1} = 0, \qquad (n = 1, 2, \ldots)$$

can be solved iteratively to give

$$p_1 = \frac{\lambda}{\mu}p_0, \quad p_2 = \frac{\lambda^2}{2!\mu^2}p_0, \quad \cdots \quad p_n = \frac{\lambda^n}{n!\mu^n}p_0, \quad \cdots.$$

Using the condition $\sum_{n=0}^{\infty} p_n = 1$ determine p_0. Find the mean steady-state population size, and compare the result with that obtained in Example 6.3.

6.19. In a simple birth process, the probability that the population is of size n at time t, given that it was n_0 at time $t = 0$, is given by

$$p_n(t) = \binom{n-1}{n_0-1} e^{-\lambda n_0 t} (1 - e^{-\lambda t})^{n-n_0}, \qquad (n \geq n_0)$$

(see Section 6.3 and Figure 6.1). Show that the probability achieves its maximum value for given n and n_0 when

$$t = (1/\lambda) \ln(n/n_0).$$

Find also the maximum value of $p_n(t)$ at this time.

6.20. In a birth and death process with equal birth and death parameters, the probability generating function is (see equation (6.24))

$$G(s, t) = \left[\frac{1 + (\lambda t - 1)(1 - s)}{1 + \lambda t (1 - s)} \right]^{n_0}.$$

Find the mean population size at time t. Show also that its variance is $2n_0 \lambda t$.

6.21. In a simple death process, the probability that a death occurs in time δt is a time-dependent parameter $\mu(t)$. The pgf $G(s, t)$ satisfies

$$\frac{\partial G}{\partial t} = \mu(t)(1 - s) \frac{\partial G}{\partial s}.$$

Show that

$$G(s, t) = [1 - e^{-\tau}(1 - s)]^{n_0},$$

where

$$\tau = \int_0^t \mu(s) \, ds.$$

Find the mean population size at time t.

In a death process it is found that the expected value of the population size at time t is given by

$$\mu(t) = \frac{n_0}{1 + \alpha t}, \qquad (t \geq 0),$$

where α is a positive constant. Estimate the corresponding death-rate $\mu(t)$.

6.22. A population process has a probability generating function $G(s, t)$, which satisfies the equation

$$e^{-t} \frac{\partial G}{\partial t} = \lambda (s - 1)^2 \frac{\partial G}{\partial s}.$$

If, at time $t = 0$, the population size is n_0, show that

$$G(s, t) = \left[\frac{1 + (1 - s)(\lambda e^t - \lambda - 1)}{1 + \lambda(1 - s)(e^t - 1)} \right]^{n_0}.$$

Find the mean population at time t, and the probability of ultimate extinction.

6.23. A population process has a probability generating function given by

$$G(s, t) = \frac{1 - \mu e^{-t}(1 - s)}{1 + \mu e^{-t}(1 - s)},$$

where μ is a parameter. Find the mean of the population size at time t, and its limit as $t \to \infty$. Expand $G(s, t)$ in powers of s, determine the probability that the population size is n at time t.

6.24. Consider a birth and death process with parameters λ and μ. In Section 6.5 it was shown that the probability generating function for the population size is given by (see equation (6.23))

$$G(s, t) = \left[\frac{\mu(1 - s) - (\mu - \lambda s)e^{-(\lambda - \mu)t}}{\lambda(1 - s) - (\mu - \lambda s)e^{-(\lambda - \mu)t}} \right]^{n_0},$$

where n_0 is the initial population size. Let

$$N(t) = N_1(t) + N_2(t) + \cdots + N_{n_0}(t),$$

where $N_i(t)$ is the number of descendants of the ith member of the original population at $t = 0$. Show that

$$\mathbf{E}[s^{N_i(t)}] = \frac{\mu(1 - s) - (\mu - \lambda s)e^{-(\lambda - \mu)t}}{\lambda(1 - s) - (\mu - \lambda s)e^{-(\lambda - \mu)t}}$$

for $1 \le i \le n_0$. If $\lambda > \mu$, what is the probability that the number of descendants from any individual of the original population becomes zero?

6.25. In a birth and death process with equal rates λ, the probability generating function is given by (see equation (6.24))

$$G(s, t) = \left[\frac{\lambda(z + t) - 1}{\lambda(z + t)} \right]^{n_0} = \left[\frac{1 + (\lambda t - 1)(1 - s)}{1 + \lambda t(1 - s)} \right]^{n_0},$$

where n_0 is the initial population size. Show that p_i, the probability that the population size is i at time t, is given by

$$p_i(t) = \sum_{m=0}^{i} \binom{n_0}{m} \binom{n_0 + i - m - 1}{i - m} \alpha(t)^m \beta(t)^{n_0 + i - m}$$

if $i \leq n_0$, and by

$$p_i(t) = \sum_{m=0}^{n_0} \binom{n_0}{m} \binom{n_0 + i - m - 1}{i - m} \alpha(t)^m \beta(t)^{n_0 + i - m}$$

if $i > n_0$, where

$$\alpha(t) = \frac{1 - \lambda t}{\lambda t}, \qquad \beta(t) = \frac{\lambda t}{1 + \lambda t}.$$

6.26. We can view the birth and death process by an alternative differencing method. Let $p_{ij}(t)$ be the conditional probability

$$p_{ij}(t) = \mathbf{P}(N(t) = j \mid N(0) = i),$$

where $N(t)$ is the random variable representing the population size at time t. Assume that the process is in the (fixed) state $N(t) = j$ at times t and $t + \delta t$ and decide how this can arise from an incremental change δt in the time. If the birth and death rates are λ_j and μ_j, explain why

$$p_{ij}(t + \delta t) = p_{ij}(t)(1 - \lambda_i \delta t - \mu_i \delta t) + \lambda_i \delta t p_{i+1,j}(t) + \mu_i \delta t p_{i-1,j}(t)$$

for $i = 1, 2, 3, \ldots$, $j = 0, 1, 2, \ldots$. Take the limit as $\delta t \to 0$, and confirm that $p_{ij}(t)$ satisfies the differential equation

$$\frac{dp_{ij}(t)}{dt} = -(\lambda_i + \mu_i)p_{ij}(t) + \lambda p_{i+1,j}(t) + \mu_i p_{i-1,j}(t).$$

How should $p_{0,j}(t)$ be interpreted?

6.27. Consider a birth and death process in which the rates are $\lambda_i = \lambda i$ and $\mu_i = \mu i$, and the initial population size is $n_0 = 1$. If

$$p_{1,j} = \mathbf{P}(N(t) = j \mid N(0) = 1),$$

it was shown in Problem 6.26 that $p_{1,j}$ satisfies

$$\frac{dp_{1,j}(t)}{dt} = -(\lambda + \mu)p_{1,j}(t) + \lambda p_{2,j}(t) + \mu p_{0,j}(t),$$

where

$$p_{0,j}(t) = \begin{cases} 0, & j > 0 \\ 1, & j = 0 \end{cases}$$

If

$$G_i(s, t) = \sum_{j=0}^{\infty} p_{ij}(t)s^j,$$

show that

$$\frac{\partial G_1(s, t)}{\partial t} = -(\lambda + \mu)G_1(s, t) + \lambda G_2(s, t) + \mu.$$

Explain why $G_2(s, t) = [G_1(s, t)]^2$ (see Section 6.5). Hence solve what is effectively an ordinary differential equation for $G_1(s, t)$, and confirm that

$$G_1(s, t) = \frac{\mu(1 - s) - (\mu - \lambda s)e^{-(\lambda - \mu)t}}{\lambda(1 - s) - (\mu - \lambda s)e^{-(\lambda - \mu)t}},$$

as in equation (6.23) with $n_0 = 1$.

6.28. In a birth and death process with parameters λ and μ, $(\mu > \lambda)$, and initial population size n_0, show that the mean time to extinction T_{n_0} is given by

$$\mathbf{E}(T_{n_0}) = n_0\mu(\mu - \lambda)^2 \int_0^\infty \frac{te^{-(\mu-\lambda)t}[\mu - \mu e^{-(\mu-\lambda)t}]^{n_0-1}}{[\mu - \lambda e^{-(\mu-\lambda)t}]^{n_0+1}} \, dt.$$

If $n_0 = 1$, using integration by parts evaluate the integral over the interval $(0, \tau)$ and then let $\tau \to \infty$ to show that

$$\mathbf{E}(T_{n_0}) = -\frac{1}{\lambda} \ln\left(\frac{\mu - \lambda}{\mu}\right).$$

6.29. A death process (see Section 6.4) has a parameter μ and the initial population size is n_0. Its probability generating function is

$$G(s, t) = [1 - e^{-\mu t}(1 - s)]^{n_0}.$$

Show that the mean time to extinction is

$$\frac{n_0}{\mu} \sum_{k=0}^{n_0-1} \frac{(-1)^k}{(k + 1)^2}\binom{n_0 - 1}{k}.$$

6.30. A colony of cells grows from a single cell without deaths. The probability that a single cell divides into two cells in a time interval δt is $\lambda + \delta t + o(\delta t)$. As in Problem 6.1, the probability generating function for the process is

$$G(s, t) = \frac{se^{-\lambda t}}{1 - (1 - e^{-\lambda t})s}.$$

By considering the probability

$$F(t) = \sum_{k=n}^\infty p_k(t),$$

that is, the probability that the population is n or greater by time t, show that the expected time T_n for the population to first reach $n \geq 2$ is given by

$$\mathbf{E}(T_n) = \frac{1}{\lambda} \sum_{k=1}^{n-1} \frac{1}{k}.$$

6.31. In a birth and death process, the population size random variable $N(t)$ grows as a simple birth process with parameter λ. No deaths occur until time τ when the whole population dies. The distribution of the random variable τ is exponential with parameter μ. The process starts with one individual at time $t = 0$. What is the probability that the population still exists at time t, namely that $\mathbf{P}[N(t) > 0]$?

What is the conditional probability

$$\mathbf{P}[N(t) = n | N(t) > 0]$$

for $n = 1, 2, \ldots$? Hence show that

$$\mathbf{P}[N(t) = n] = e^{(\lambda+\mu)t}(1 - e^{\lambda t})^{n-1}.$$

Construct the probability generating function of this distribution, and find the mean population size at time t.

6.32. The following birth and death process is a possible model for the random opening and closing of membrane channels in neuroscience (see Tuckwell, 1989). Let $p_n(t)$ be the probability that the population size is n at time t where $0 \le n \le a$: in this process, the population size cannot exceed a. In the notation of equation (6.25),

$$\frac{dp_0(t)}{dt} = -\lambda_0 p_0(t) + \mu_1 p_1(t),$$

$$\frac{dp_n(t)}{dt} = \lambda_{n-1} p_{n-1}(t) - (\lambda_n + \mu_n) p_n(t) + \mu_{n+1} p_{n+1}(t), \quad (1 \le n \le a-1),$$

$$\frac{dp_a(t)}{dt} = \lambda_{a-1} p_{a-1} - \mu_a p_a,$$

where

$$\lambda_n = (a - n)\lambda, \qquad \mu_n = n\mu \quad (n = 0, 1, 2, \ldots a).$$

Show that the probability generating function

$$G(s, t) = \sum_{n=0}^{a} p_n(t) s^n$$

satisfies the partial differential equation

$$\frac{\partial G(s, t)}{\partial t} = \lambda a(s - 1) G(s, t) + [\mu + (\lambda - \mu)s - \lambda s^2]\frac{\partial G(s, t)}{\partial s}.$$

If the initial population size is $a_0 \le a$ at time $t = 0$, verify that

$$G(s, t) = (\lambda + \mu)^{-a}[\mu + \lambda s + \mu(s - 1)e^{-(\lambda+\mu)t}]^{a_0}$$
$$\times [\mu + \lambda s - \lambda(s - 1)e^{-(\lambda+\mu)t}]^{a-a_0}.$$

Find the mean population size from this generating function and confirm that it approaches $a\lambda/(\lambda + \mu)$ as $t \to \infty$.

7
Queues

7.1 Introduction

Queues appear in many aspects of life. Some are clearly visible, as in the queues
at supermarket check-out tills: others may be less obvious, as in call-stacking at
airline information telephone lines, or in hospital waiting lists. In the latter, an
individual waiting may have no idea how many persons are in front of him or
her in the hospital appointments list. Generally, we are interested in the long-
term behaviour of queues for future planning purposes – does a queue increase
with time or does it have a steady state, and if it does have a steady state then,
on average, how many individuals are there in the queue and what is the mean
waiting time?

We can introduce some of the likely hypotheses behind queues by looking at
a familiar example from banking. Consider a busy city centre cash dispensing
machine (often called an ATM) outside a bank. As an observer what do we see?
Individuals, or *customers* as they are known in queueing processes, approach the
dispenser, or *server* as it is known. If there is no one using the machine, then
the customer inserts a cash card obtains cash or transacts other business. There
is a period of time when the customer is being served. On the other hand, if
there is already someone at the till then the assumption is that the customer and
succeeding ones form a queue and wait their turn.

We have to make assumptions about both the *service time* and the *customer arrivals*. We could assume that service takes a *fixed* time (discussed in Section 7.5)
or, more likely, that the service times are random variables, which are independent
for each customer. They could perhaps have a negative exponential density

$$f(t) = \mu e^{-\mu t}, \qquad t \geq 0.$$

This means that the customer spends on average time $1/\mu$ at the dispenser or
being served.

For most queues, the assumption is that customers are served in the order of
arrival at the end of the queue. The basis of this is the '*first come, first served*' rule.
However, if queueing is not taking place, then the next customer could be chosen
at random from those waiting. This could be the case where the 'customers' are

components in a manufacturing process, which are being delivered, stored until required and then chosen at random for the next process. In this application, the arrivals could be regular and predictable. The simplest assumption with arrivals at a queue is that they have a Poisson distribution with parameter λt, so that λ is the average number of arrivals per unit time. From our discussion of Poisson processes in Chapter 5, this implies that the probability of an arrival in the time interval $(t, t + \delta t)$ is $\lambda \delta t$ regardless of what happened before time t. And, of course, the probability of two or more arrivals in the time interval is negligible. As with the service time, the density function of the time interval between arrivals is negative exponential:

$$\lambda e^{-\lambda t}.$$

In practice, however, there may be busy periods. Arrivals at our cash dispenser are unlikely to be uniformly distributed over 24 hour operation: there will be more customers at peak periods during the day. If a queue lengthens too much at peak times to the point of deterring customers, the bank might wish to install a second cash dispenser. Hence, a queue with a *single server* now becomes a *two-server queue*. Generalizing this, we can consider n-server queues, which are typical queueing processes that occur in supermarkets, the customer having to choose which check-out appears to have the quickest service time. Alternatively, the n servers could be approached through a single queue with customers being directed to the next free server.

There many aspects of queues that can be considered but not all of them will be investigated here. Customers could baulk at queueing if there are too many people ahead of them. Queues can have a maximum length: when the queue reaches a prescribed length further customers are turned away.

Generally, we are not concerned with how queues begin when, say, a bank opens, but how they develop over time as the day progresses. Do the queues become longer, and if not what is the average length of the queue? Often, therefore, we look at the underlying *stationary process* associated with the queue. This effectively takes time variations out of the problem with the result that differential-difference equations for recurrent probabilities of the Markov process reduce to difference equations. Queueing processes are related to the Poisson and birth and death processes of Chapters 4 and 5 when they are Markov, and the construction of the difference equations follows from similar arguments and hypotheses.

7.2 The single-server queue

We will now put together a simple model of a queue that has close affinity with the Poisson and birth and death processes of Chapters 4 and 5. Let the random variable $N(t)$ denote the number of individuals in the queue, including the person being served. We assume that there is a single counter or server with an orderly forming queue served on a first come, first served basis. As before we write

$$\mathbf{P}(N(t) = n) = p_n(t),$$

which is the probability that there are n individuals in the queue at time t. We now look at the probability that there are n persons in the queue at time $t + \delta t$. We have to specify the probability that an individual arrives at the end of the queue in the time interval δt. Probably the simplest assumption is that the probability of an arrival is $\lambda \delta t$, where λ is a constant. In other words, the arrivals form a Poisson process with parameter λt. We must also make an assumption about the service time of any customer at the counter. Assume that these times form random time intervals with exponential density given by

$$f(t) = \mu e^{-\mu t}, \qquad t \geq 0,$$

these times being measured from the arrival of a customer at the counter. The question is: what is the probability that a customer leaves the counter in time δt? If T is a random variable representing the service time (the time spent at the counter by the customer), then

$$\mathbf{P}(t \leq T \leq t + \delta t | T \geq t)$$

represents the probability that the service will be completed in the interval $(t, t + \delta t)$, given that it is still in progress at time t. For this conditional probability:

$$\mathbf{P}(t \leq T \leq t + \delta t | T \geq t) = \frac{\mathbf{P}(t \leq T \leq t + \delta t)}{\mathbf{P}(T \geq t)} = \frac{\int_t^{t+\delta t} \mu e^{-\mu s} \, ds}{\int_t^{\infty} \mu e^{-\mu s} \, ds}$$

$$= \frac{[-e^{-\mu s}]_t^{t+\delta t}}{[-e^{-\mu s}]_t^{\infty}} = \frac{-e^{-\mu(t+\delta t)} + e^{-\mu t}}{e^{-\mu t}}$$

$$= 1 - e^{-\mu \delta t}$$

$$\approx 1 - (1 - \mu \delta t) = \mu \delta t.$$

for small δt, using a two-term Taylor expansion for $e^{-\mu \delta t}$. As we would expect with this exponential distribution, the probability of the service being completed in the interval $(t, t + \delta t)$ is independent of the current state of the service.

By the law of total probability (Section 1.3) we have

$$p_n(t + \delta t) = \lambda \delta t p_{n-1}(t) + \mu \delta t p_{n+1} + (1 - \lambda \delta t - \mu \delta t) p_n(t) + o(\delta t),$$

$$p_0(t + \delta t) = \mu \delta t p_1(t) + (1 - \lambda \delta t) p_0(t),$$

where multiple arrivals are assumed to be negligible. Thus

$$\frac{p_n(t + \delta t) - p_n(t)}{\delta t} = \lambda p_{n-1}(t) + \mu p_{n+1}(t) - (\lambda + \mu) p_n(t) + o(1),$$

$$\frac{p_0(t + \delta t) - p_0(t)}{\delta t} = \mu p_1(t) - \lambda p_0(t) + o(1).$$

Let $\delta t \to 0$ so that

$$\left.\begin{array}{r}
\dfrac{\mathrm{d}p_n(t)}{\mathrm{d}t} = \lambda p_{n-1}(t) + \mu p_{n+1}(t) - (\lambda + \mu)p_n(t), \qquad (n = 1, 2, \ldots) \\[2mm]
\dfrac{\mathrm{d}p_0(t)}{\mathrm{d}t} = \mu p_1(t) - \lambda p_0(t)
\end{array}\right\}$$

(7.1)

These equations are awkward to solve by the probability generating function method. However, it is possible to solve the equations using this method, and the result, after considerable analysis, leads to the formulas for probabilities in terms of special functions, known as *Bessel functions*. Fortunately, we can obtain considerable information about queues by looking at the underlying *stationary process* in equation (7.1).

7.3 The stationary process

In the stationary process, we look at the long-term behaviour of the queue if it has a limiting state. Not all queues do: if the rate of arrivals is large compared with the service time then we might expect the queue to continue to grow without bound as time progresses. For the moment, let us assume that there is a limiting state, that $\lambda < \mu$ (otherwise we would expect the queue just to increase in length with time) and that we can write

$$\lim_{t \to \infty} p_n(t) = p_n.$$

The probabilities now approach a *constant* distribution that does not depend on time. In this case it is reasonable to assume that

$$\lim_{t \to \infty} \frac{\mathrm{d}p_n(t)}{\mathrm{d}t} = 0.$$

If the limits turn out to be not justifiable then we might expect some inconsistency to occur in the limiting form of the equations or in their solution, which is a backward justification.

Equations (7.1) now become

$$\left.\begin{array}{r}
\lambda p_{n-1} + \mu p_{n+1} - (\lambda + \mu)p_n = 0, \qquad (n = 1, 2, \ldots) \\[2mm]
\mu p_1 - \lambda p_0 = 0
\end{array}\right\}$$

(7.2)

The general equation in (7.2) is a second-order *difference equation*. It is easy to solve iteratively since

$$p_1 = \frac{\lambda}{\mu} p_0$$

and, with $n = 1$,

$$p_2 = \frac{1}{\mu}[(\lambda + \mu)p_1 - \lambda p_0] = \frac{1}{\mu}[(\lambda + \mu)\frac{\lambda}{\mu} p_0 - \lambda p_0] = \left(\frac{\lambda}{\mu}\right)^2 p_0.$$

The next iterate is

$$p_3 = \left(\frac{\lambda}{\mu}\right)^3 p_0,$$

and, in general,

$$p_n = \left(\frac{\lambda}{\mu}\right)^n p_0. \tag{7.3}$$

Alternatively, the second-order difference equation can be solved using the characteristic equation method (see Section 2.2). The characteristic equation of (7.2) is

$$\mu m^2 - (\lambda + \mu)m + \lambda = 0, \qquad \text{or} \qquad (\mu m - \lambda)(m - 1) = 0.$$

The roots are $m_1 = 1, m_2 = \lambda/\mu$. If $\lambda \neq \mu$, then the general solution is

$$p_n = A + B\left(\frac{\lambda}{\mu}\right)^n; \tag{7.4}$$

if $\lambda = \mu$, it is

$$p_n = C + Dn,$$

where A, B, C and D are constants. Since $\mu p_1 = \lambda p_0$, it follows that $A = 0$ or $D = 0$. Let $B = C = p_0$: the result is (7.3) again.

Let $\rho = \lambda/\mu$ for future reference: ρ is known as the *traffic density*. Generally, the traffic density is defined as

$$\rho = \frac{\mathbf{E}(S)}{\mathbf{E}(I)},$$

where S and I are independent random variables of the service and inter-arrival times respectively. When S and I are both negative exponentially distributed, then $\mathbf{E}(S) = 1/\mu$ and $\mathbf{E}(I) = 1/\lambda$. We still require the value of p_0. However, since $\sum_{n=0}^{\infty} p_n = 1$, it follows that

$$\left(\sum_{n=0}^{\infty} \rho^n\right) p_0 = 1,$$

but the geometric series on the right will only *converge* if $|\rho| < 1$. Its sum is then given by

$$\sum_{n=0}^{\infty} \rho^n = \frac{1}{1 - \rho}.$$

Hence

$$p_0 = 1 - \rho,$$

and

$$p_n = (1 - \rho)\rho^n, \qquad (n = 0, 1, 2, \ldots), \tag{7.5}$$

which is a geometric distribution (see Section 1.7).

If $\rho \geq 1$ then the series does not converge and $\{p_n\}$ cannot represent a probability distribution since the series $\sum_{n=0}^{\infty} p_n$ *diverges*. In this case, the queue simply increases in length to infinity as $t \to \infty$, whatever its initial length. On the other hand, if $\rho < 1$, then $\lambda < \mu$ and the rate of arrivals is less than the average service time with the result that a steady state should be achieved, as we might expect.

There are various items of further information about the single-server queue that we can deduce from the distribution $\{p_n\}$ assuming $\rho < 1$.

(a) Server free

The probability that the server is free when a customer arrives is

$$p_0 = 1 - \rho.$$

If the service time parameter is adjustable, for example, then the question might arise as to what might be an acceptable figure for p_0.

(b) Length of queue

The mean length of the queue (including the person being served) is, if N represents a random variable of the number in the queue,

$$\mathbf{E}(N) = \sum_{n=1}^{\infty} n p_n = \sum_{n=1}^{\infty} n(1 - \rho)\rho^n = \sum_{n=1}^{\infty} n\rho^n - \sum_{n=1}^{\infty} n\rho^{n+1}$$

$$= \sum_{n=1}^{\infty} n\rho^n - \sum_{n=1}^{\infty} (n - 1)\rho^n = \sum_{n=1}^{\infty} \rho^n = \frac{\rho}{1 - \rho},$$

or the result could be obtained directly from the mean of the geometric distribution in (7.5). Hence, if $\rho = \frac{3}{4}$, then a customer might expect one person being served and two queueing. If, for example, the service time parameter μ can be varied (customers are served more efficiently perhaps), then $\mathbf{E}(N) = 2$ could be set at a level acceptable to arriving customers. For example, setting $\mathbf{E}(N) = 2$ implies that $\rho = \frac{2}{3}$, or $\mu = \frac{3}{2}\lambda$.

(c) Waiting time

How long will a customer expect to wait for service on arrival at the back of the queue, and how long will the customer expect to queue and be served? A customer arrives and finds that there are n individuals ahead, including the person being served. Let T_i be the random variable representing the time for the service of customer i. We are interested in the random variable

$$S_n = T_1 + T_2 + \cdots + T_n,$$

the sum of the random variables of service times of the first n queueing customers. These random variables are independent, each with an exponential density function with parameter μ, that is,

$$f(t) = \mu e^{-\mu t}.$$

The moment generating function of the random variable T_i is

$$M_{T_i}(t) = \mathbf{E}(e^{tT_i}) = \int_0^\infty e^{ts} f(s)\, ds = \mu \int_0^\infty e^{ts} e^{-\mu s}\, ds,$$

$$= \mu \int_0^\infty e^{-(\mu - t)s}\, ds,$$

$$= \frac{\mu}{\mu - t}, \qquad (t < \mu).$$

We need the result that the moment generating function of the sum

$$S_n = T_1 + T_2 + \cdots + T_n,$$

is

$$M_{S_n}(t) = \mathbf{E}(e^{tS_n}) = \mathbf{E}(e^{t(T_1 + T_2 + \cdots + T_n)})$$

$$= \mathbf{E}(e^{tT_1})\mathbf{E}(e^{tT_2}) \cdots \mathbf{E}(e^{tT_n})$$

$$= M_{T_1}(t) M_{T_2}(t) \cdots M_{T_n}(t)$$

$$= \left(\frac{\mu}{\mu - t}\right)^n$$

(see Section 1.9). This is the moment generating function of the gamma density function with parameters μ and n. Its density is

$$f_{S_n}(t) = \frac{\mu^n}{(n-1)!} t^{n-1} e^{-\mu t}$$

(see Section 1.8).

Let

$$S = \lim_{n \to \infty} S_n,$$

which is the limiting waiting time, assuming that it exists. Using the law of total probability, the probability that S is greater than a time t is given by

$$\mathbf{P}(S > t) = \sum_{n=1}^\infty \mathbf{P}(S_n > t | n) p_n.$$

Hence

$$\mathbf{P}(S > t) = \sum_{n=1}^{\infty} p_n \int_t^{\infty} f_{S_n}(s)\, ds,$$

$$= \sum_{n=1}^{\infty} (1-\rho)\rho^n \int_t^{\infty} \frac{\mu^n}{(n-1)!} s^{n-1} e^{-\mu s}\, ds,$$

$$= (1-\rho)\mu\rho \int_t^{\infty} e^{-\mu s} \sum_{n=0}^{\infty} \frac{(\mu\rho s)^n}{n!}\, ds,$$

$$= (1-\rho)\mu\rho \int_t^{\infty} e^{-\mu(1-\rho)s}\, ds,$$

$$= \rho e^{-\mu(1-\rho)t}.$$

The associated density function is

$$g(t) = \frac{d}{dt}[1 - \rho e^{-\mu(1-\rho)t}] = \rho\mu(1-\rho)e^{-\mu(1-\rho)t}.$$

Finally, the expected value of S is

$$\mathbf{E}(S) = \int_0^{\infty} t g(t)\, dt,$$

$$= \int_0^{\infty} t\rho\mu(1-\rho)e^{-\mu(1-\rho)t}\, dt,$$

$$= \rho\mu(1-\rho)\cdot\frac{1}{\mu^2(1-\rho)^2} \int_0^{\infty} s e^{-s}\, ds$$

$$= \frac{\rho}{\mu(1-\rho)},$$

which turns out to be the expected value $1/\mu$ of the service time T multiplied by the expected length of the queue $\rho/(1-\rho)$ from (ii) above. Hence in this case, it occurs that

$$\mathbf{E}(S) = \mathbf{E}(N)\mathbf{E}(T).$$

This is a lengthy argument: the difficulty was mainly caused by the problem of finding the density function of the sum of a set of random variables.

(d) Busy periods

If $N(t)$ is the random number representing the number of individuals in a single-server queue at time t in the non-stationary process, then the queue length with $\rho = \lambda/\mu < 1$ might look as in Figure 7.1. Being now a display of a continuous time process, the numbers in the queue will change up or down at varying times. There will be periods where the server is free (called *slack periods*), periods of lengths

$$s_1 = t_1, \quad s_2 = t_3 - t_2, \quad s_3 = t_5 - t_4, \ldots,$$

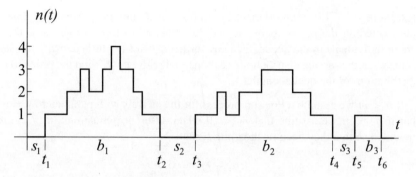

Fig. 7.1 *An example of a queue where the number in the queue $n(t)$ is plotted against time in the non-stationary process: the queue starts with no customers and $\{s_i\}$ are the slack periods and $\{b_i\}$ are the busy periods.*

and periods when the server is busy, namely

$$b_1 = t_2 - t_1, \quad b_2 = t_4 - t_3, \quad b_3 = t_6 - t_5, \ldots.$$

The times $t_1, t_3, t_5, \ldots$ are the times when a new customer arrives when the server is free, and $t_2, t_4, t_6, \ldots$ are the times at which the server becomes free. The periods denoted by $b_1, b_2, \ldots$ are known as *busy periods*. A question whose answer is of interest is: what is the expected length of a busy period? Here we present an informal argument that points to the result.

Suppose that the server is free, as at time $t = t_2$. Then, since the arrivals form a Poisson process with parameter λ, the expected time until the next customer arrives is $1/\lambda$, since the density function is $\lambda e^{-\lambda t}$. It seems a reasonable intuitive result that the average lengths of a large number, n, of slack periods should approach $1/\lambda$, that is,

$$\lim_{n \to \infty} \left[\frac{\sum_{i=1}^{n} s_i}{n} \right] = \frac{1}{\lambda}.$$

In the stationary process, the probability that the server is free is p_0 and that the server is busy is $1 - p_0$. Hence, the ratio of the average lengths of the slack periods to that of the busy periods is $p_0/(1 - p_0)$. Hence

$$\lim_{n \to \infty} \left[\frac{\sum_{i=1}^{n} s_i}{n} \cdot \frac{n}{\sum_{i=1}^{n} b_i} \right] = \frac{p_0}{1 - p_0} = \frac{1 - \rho}{\rho}.$$

Since the limit of the product is the product of the limits, it follows using the limit above that

$$\lim_{n \to \infty} \left[\frac{1}{n} \sum_{i=1}^{n} b_i \right] = \frac{1}{\lambda} \frac{\rho}{1 - \rho} = \frac{1}{\mu - \lambda},$$

which is the mean length of the busy periods.

Example 7.1 (The baulked queue). In this queue, not more than $m \geq 2$ people (including the person being served) are allowed to form a queue. If there are m individuals in the queue, then any further arrivals are turned away. It is assumed that the service distribution is exponential with rate μ. Find the probability distribution for the queue length.

If $n < m$ we assume a Poisson process for the arrivals with parameter $\lambda_n = \lambda$, but if $n \geq m$ we assume that $\lambda_n = 0$. The difference equations in (7.2) are modified so that they become the finite system

$$\mu p_1 - \lambda p_0 = 0,$$

$$\lambda p_{n-1} + \mu p_{n+1} - (\lambda + \mu) p_n = 0, \qquad (1 \leq n \leq m - 1),$$

$$\lambda p_{m-1} - \mu p_m = 0.$$

The general solution is (see equation (7.4))

$$p_n = A + B\rho^n, \qquad (\rho \neq 1).$$

For the baulked queue there is no restriction on ρ, but $\rho = 1$ must be treated separately.

The *boundary conditions* imply

$$\mu(A + B\rho) - \lambda(A + B) = 0,$$

$$\lambda(A + B\rho^{m-1}) - \mu(A + B\rho^m) = 0,$$

or,

$$A(1 - \rho) = 0$$

in both cases. Since $\rho \neq 1$, we conclude that $A = 0$. Hence

$$p_n = B\rho^n, \qquad (n = 0, 1, 2, \ldots, m).$$

This distribution must satisfy $\sum_{n=0}^{m} p_n = 1$. Hence

$$B \sum_{n=0}^{m} \rho^n = 1,$$

or, after summing the geometric series,

$$B \frac{1 - \rho^{m+1}}{1 - \rho} = 1.$$

Hence, the probability that there are n individuals in the queue is given by

$$p_n = \frac{\rho^n (1 - \rho)}{(1 - \rho^{m+1})}, \qquad (n = 0, 1, 2, \ldots, m) \quad (\rho \neq 1).$$

If $\rho = 1$, then the general solution for p_n is

$$p_n = A + Bn,$$

and the boundary condition $p_1 - p_1 = 0$ implies $B = 0$. Hence, $p_n = A$ ($n = 0, 1, 2, \ldots m$), but since $\sum_{n=0}^{m} p_n = 1$, it follows that $A = 1/(m + 1)$. Finally,

$$p_n = \frac{1}{m+1}, \qquad (n = 0, 1, 2, \ldots m),$$

which is a discrete uniform distribution. □

7.4 Queues with multiple servers

In many practical applications of queueing models, there is more than one server; as in banks, supermarkets and hospital admissions. Usually in supermarkets, shoppers at check-outs, choose a till with the shortest queue or, more accurately, a queue with the smallest number of likely purchases. On the other hand, banks and post offices, for example, often guide arrivals into a single queue, and then direct them to tills or counters as they become free. It is the second type of queue that we will investigate. A plan of such a scheme is shown in Figure 7.2.

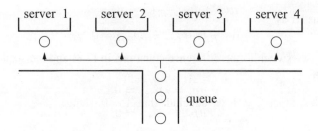

Fig. 7.2 *A queue with 4 servers and 3 individuals waiting.*

Suppose that the queue faces r servers. As for the single-server queue, assume that the arrivals form a Poisson process with parameter λt, and that the service time at each counter has exponential density with rate μ. Let $p_n(t)$ be the probability that there are n people in the queue at time t, including those being served. The queue length is governed by different equations depending on whether $n < r$ or $n \geq r$. If $n \geq r$ then all counters will be occupied by a customer, but some will be free if $n < r$. If $n < r$ the probability of a counter becoming free in time δt will be $n\mu\delta t$: if $n \geq r$ the probability will be $r\mu\delta t$. Proceeding as in Section 6.2, the equations or $p_n(t)$ are, for $r \geq 2$,

$$\frac{\mathrm{d}p_0(t)}{\mathrm{d}t} = \mu p_1(t) - \lambda p_0(t),$$

$$\frac{dp_n(t)}{dt} = \lambda p_{n-1}(t) + (n+1)\mu p_{n+1}(t) - (\lambda + n\mu)p_n(t), \qquad 1 \le n < r,$$

$$\frac{dp_n(t)}{dt} = \lambda p_{n-1}(t) + r\mu p_{n+1}(t) - (\lambda + r\mu)p_n(t), \qquad n \ge r.$$

The corresponding stationary process for the queue with r servers is

$$\mu p_1 - \lambda p_0 = 0, \tag{7.6}$$

$$\lambda p_{n-1} + (n+1)\mu p_{n+1} - (\lambda + n\mu)p_n = 0, \qquad 1 \le n < r, \tag{7.7}$$

$$\lambda p_{n-1} + r\mu p_{n+1} - (\lambda + r\mu)p_n = 0, \qquad n \ge r. \tag{7.8}$$

This is not now a set of constant-coefficient difference equations, and the characteristic equation method of solution no longer works. However, it is easy to solve the equations iteratively. For $n = 1$ in equation (7.7), we have

$$\lambda p_0 + 2\mu p_2 - (\lambda + \mu)p_1 = 0,$$

or, using (7.6),

$$2\mu p_2 - \lambda p_1 = 0.$$

If $n < r$ we can repeat this procedure for $n = 2$ giving

$$3\mu p_3 - \lambda p_2 = 0$$

and so on as far as

$$(n-1)\mu p_{r-1} - \lambda p_{r-2} = 0.$$

For $n \ge r$, we switch to the difference equations in (7.8) and the sequence then continues as

$$r\mu p_r = \lambda p_{r-1} = 0,$$

$$r\mu p_{r+1} - \lambda p_r = 0,$$

etc. To summarize: the set of difference equations (7.6)–(7.8) reduces to the equivalent set

$$n\mu p_n = \lambda p_{n-1}, \qquad 1 \le n < r, \tag{7.9}$$

$$r\mu p_n = \lambda p_{n-1}, \qquad n \ge r. \tag{7.10}$$

Starting with $n = 1$,

$$p_1 = \frac{\lambda}{\mu} p_0 = \rho p_0,$$

where $\rho = \lambda/\mu$. Then for $n = 2$,

$$p_2 = \frac{\rho}{2} p_1 = \frac{\rho^2}{2} p_0,$$

and so on. Thus, for $n < r$,

$$p_n = \frac{\rho^n}{n!} p_0,$$

and for $n \geq r$

$$p_n = \frac{\rho^n}{r^{n-r}r!}p_0.$$

Since $\sum_{n=0}^{\infty} p_n = 1$, we can determine p_0 from

$$p_0 \left[\sum_{n=0}^{r-1} \frac{\rho^n}{n!} + \frac{r^r}{r!} \sum_{n=r}^{\infty} \left(\frac{\rho}{r}\right)^n \right] = 1,$$

in which the infinite geometric series converges if $\rho < r$. As we would expect in the multiple-server queue, it is possible for the arrival rate λ to be greater than the service rate μ. Summing the second geometric series, we obtain

$$p_0 = 1 \bigg/ \left[\sum_{n=0}^{r-1} \frac{\rho^n}{n!} + \frac{\rho^r}{(r-\rho)(r-1)!} \right]. \tag{7.11}$$

If N is the random variable of the queue length n, then the expected queue length, *excluding* those being served, is

$$\mathbf{E}(N) = \sum_{n=r+1}^{\infty} (n-r)p_n,$$

$$= \frac{p_0 \rho^r}{r!} \sum_{n=r+1}^{\infty} (n-r) \left(\frac{\rho}{r}\right)^{n-r}.$$

We need to sum the series. Let

$$R = \sum_{n=r+1}^{\infty} (n-r) \left(\frac{\rho}{r}\right)^{n-r} = \frac{\rho}{r} + 2\left(\frac{\rho}{r}\right)^2 + 3\left(\frac{\rho}{r}\right)^3 + \cdots.$$

Multiply both sides by ρ/r and subtract the new series from the series for R. Hence

$$R\left(1 - \frac{\rho}{r}\right) = \left(\frac{\rho}{r}\right) + \left(\frac{\rho}{r}\right)^2 + \left(\frac{\rho}{r}\right)^3 + \cdots = \frac{\rho}{r-\rho}$$

using the formula for the sum of a geometric series. Hence

$$R = \frac{\rho r}{(r-\rho)^2},$$

and

$$\mathbf{E}(N) = \frac{p_0 \rho^{r+1}}{(r-1)!(r-\rho)^2}, \tag{7.12}$$

where p_0 is given by equation (7.11).

This formula gives the expected length of the queue including those actually waiting for service. The expected length including those being served can be found in the solution to Problem 7.9.

For a queue with r servers, $\rho/r = \lambda/r\mu$ is a measure of the traffic density of the queue, and it is this parameter which must be less than 1 for the expected length of the queue to remain finite.

Example 7.2. A bank has two tellers but one is inexperienced and takes longer than average to complete the service required by a customer. If customers arrive as a Poisson process with parameter λt, and the service times are independent and exponentially distributed with rates μ_1 and μ_2, find the probability that the queue contains, in the stationary process, n individuals excluding those being served. What condition must λ, μ_1 and μ_2 satisfy for the queue to remain bounded in length, and find its expected length in this case.

When one counter is occupied, the probability that it becomes free is either $\mu_1 \delta t$ or $\mu_2 \delta t$ in a small time interval δt with the choice of each counter being equally likely. Use the law of total probability (see Section 1.3) as follows. Let B, A_1 and A_2 be the events

$$B = \text{exactly one counter becomes free in time } \delta t,$$

$$A_1 = \text{counter 1 occupied}, \qquad A_2 = \text{counter 2 occupied}.$$

Now

$$\mathbf{P}(B|A_1) = \mu_1 \delta t, \qquad \mathbf{P}(B|A_2) = \mu_2 \delta t, \qquad \mathbf{P}(A_1) = \mathbf{P}(A_2) = \frac{1}{2}.$$

Hence

$$\mathbf{P}(B) = \mathbf{P}(B|A_1)\mathbf{P}(A_1) + \mathbf{P}(B|A_2)\mathbf{P}(A_2) = \frac{1}{2}(\mu_1 + \mu_2).$$

When both counters are occupied, the probability that a server becomes free is

$$\mathbf{P}(\text{counter 1 becomes free} \quad or \quad \text{counter 2 becomes free}) = (\mu_1 + \mu_2)\delta t$$

in time δt since the events are mutually exclusive. By comparison with equations (7.6)–(7.8), the difference equations are

$$\frac{1}{2}(\mu_1 + \mu_2)p_1 - \lambda p_0 = 0,$$

$$(\mu_1 + \mu_2)p_2 + \lambda p_0 - [\lambda + \frac{1}{2}(\mu_1 + \mu_2)]p_1 = 0,$$

$$(\mu_1 + \mu_2)p_{n+1} + \lambda p_{n-1} - [\lambda + (\mu_1 + \mu_2)]p_n = 0, \qquad N \geq 2.$$

If we let $\mu = \frac{1}{2}(\mu_1 + \mu_2)$, then this is equivalent to the two-server queue, each with exponential distrbution with the same rate μ. Hence, the required answers can be read off from the results for the multi-server queue presented immediately before this example. Thus, provided $\lambda < 2\mu$,

$$p_1 = \rho p_0,$$

$$p_n = \frac{\rho^n}{2^{n-2}2!}, \qquad n \geq 2,$$

where $\rho = \lambda/\mu$, and

$$p_0 = 1 \bigg/ \left[1 + \rho + \frac{\rho^2}{2 - \rho} \right],$$

$$= \frac{2 - \rho}{2 + \rho}.$$

In this case, the expected length of the queue is (excluding those being served)

$$\mathbf{E}(N) = \frac{p_0 \rho^3}{(2 - \rho)^2} = \frac{\rho^3}{4 - \rho^2}.$$

□

Example 7.3. Customers arrive in a bank with two counters at a rate $2\lambda \delta t$ in any small time interval δt. Service at either counter is exponentially distributed with rate μ. Which of the following schemes leads to the shorter overall queue length?

(a) A single queue feeding two servers.
(b) Two separate single-server queues with the assumptions that customers arrive at each queue with parameter $\lambda \delta t$, and choose servers at random.

(a) This is the two-server queue with rate 2λ. Hence, from equation (7.11) with $r = 2$,

$$p_0 = \frac{2 - \rho_1}{(1 + \rho_1)(2 - \rho_1) + \rho_1^2},$$

where $\rho_1 = 2\lambda/\mu$. From (7.12), the expected queue length will be, except for those being served,

$$\mathbf{E}_a(N) = \frac{p_0 \rho_1^3}{(2 - \rho_1)^2} = \frac{\rho_1^3}{(4 - \rho_1^2)} = \frac{(2\lambda/\mu)^3}{4 - (2\lambda/\mu)^2} = \frac{2\rho^3}{1 - \rho^2}, \qquad (7.13)$$

where $\rho = \lambda/\mu$.

(b) Note that customers do not exert a choice in this queueing scheme: it is not quite the same problem as customers arriving and choosing the shorter queue. For the scheme as described, the expected length of the two single-server queues is twice that of one, and is, excluding those being served,

$$\mathbf{E}_b(N) = 2 \sum_{n=2}^{\infty} (n - 1) p_n,$$

where $p_n = (1 - \rho)\rho^n$ from (7.4). Hence

$$\mathbf{E}_b(N) = 2 \left[\frac{\rho}{1 - \rho} - p_1 - (1 - p_0 - p_1) \right]$$

using the results

$$\sum_{n=1}^{\infty} n p_n = \frac{\rho}{1-\rho}, \qquad \sum_{n=0}^{\infty} p_n = 1.$$

Hence, since $p_0 = 1 - \rho$ and $p_1 = (1-\rho)\rho$

$$\mathbf{E}_b(N) = \frac{2\rho^2}{1-\rho}. \tag{7.14}$$

We must compare $\mathbf{E}_a(N)$ and $\mathbf{E}_b(N)$ given by equations (7.13) and (7.14). We can find any points where they are equal. Thus, $\mathbf{E}_a(N) = \mathbf{E}_b(N)$ if

$$\frac{2\rho^3}{1-\rho^2} = \frac{2\rho^2}{1-\rho}, \qquad 0 \le \rho < 1,$$

which only has the solution $\rho = 0$. A spot check at, say, $\rho = \frac{1}{2}$ gives

$$\mathbf{E}_a(N) = \tfrac{1}{3} \quad \text{and} \quad \mathbf{E}_b(N) = 1.$$

Hence

$$\mathbf{E}_b(N) > \mathbf{E}_a(N)$$

for all ρ such that $0 < \rho < 1$. We conclude in this model that the two single-server queues have a longer expected total queue length than the two-server queue, which might perhaps be expected. Comparative graphs of $\mathbf{E}_a(N)$ and $\mathbf{E}_b(N)$ versus r are shown in Figure 7.3.

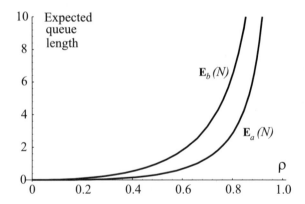

Fig. 7.3 *Comparison of expected queue lengths for the two-server queue and the two single-server queues against {arrival rate}/{service rate} ratio.*

□

7.5 Queues with fixed service times

In some queues the service time is a fixed value, say τ, since the customers all go through the same service. This could arise, for example, at a bank ATM where cash is dispensed. Suppose that the density function of the interval between arrivals is negative exponential with parameter λ, namely $\lambda e^{-\lambda t}$; that is, a Poisson arrival rate.

This queueing problem requires a different approach. As before, let $p_n(t)$ be the probability that there are n persons in the queue at time t including the person being served. We consider time intervals of duration τ, the service time, and start by asking: what is the probability $p_0(t + \tau)$ that the queue has no customers at time $t + \tau$? This can arise if there was either no one in the queue at time t, or one person at time t who must have been served and left. The probability that this occurs is $p_0(t) + p_1(t)$. In addtion, there must have been no arrivals. Since the inter-arrival times have a negative exponential distribution, the probability of no arrivals in any time interval of duration τ is $e^{-\lambda \tau}$ (see Section 5.6, and remember its no memory property). Hence

$$p_0(t + \tau) = [p_0(t) + p_1(t)]e^{-\lambda \tau}.$$

We now generalize this equation. Suppose that there are n persons in the queue at time $t + \tau$. This could have arisen in any of the ways listed below:

Number in queue at time t	Number of arrivals	Probability
0	n	$p_0(t)e^{-\lambda \tau}(\lambda \tau)^n/n!$
1	n	$p_1(t)e^{-\lambda \tau}(\lambda \tau)^n/n!$
2	$n - 1$	$p_2(t)e^{-\lambda \tau}(\lambda \tau)^{n-1}/(n - 1)!$
...	...	...
n	1	$p_n(t)e^{-\lambda \tau}\lambda \tau/1!$
$n + 1$	0	$p_{n+1}(t)e^{-\lambda \tau}$

In the table, the probability of r arrivals in time τ is $e^{-\lambda \tau}(\lambda \tau)^r/r!$. Thus, $p_n(t+\tau)$ is the sum of the probabilities in the third column:

$$p_n(t + \tau) = e^{-\lambda \tau}\left[p_0(t)\frac{(\lambda \tau)^n}{m!} + \sum_{r=1}^{n+1} \frac{p_r(t)(\lambda \tau)^{n+1-r}}{(n + 1 - r)!} \right] \quad (n = 0, 1, \ldots)$$

Suppose now that we just consider the long-term behaviour of the queue by assuming that

$$\lim_{t \to \infty} p_n(t) = \lim_{t \to \infty} p_n(t + \tau) = p_n,$$

and that the limits exist. Then the sequence $\{p_n\}$ satisfies the difference equations

$$p_n = e^{-\lambda \tau}\left[p_0\frac{(\lambda \tau)^n}{m!} + \sum_{r=1}^{n+1} \frac{p_r(\lambda \tau)^{n+1-r}}{(n + 1 - r)!} \right] \quad (n = 0, 1, \ldots).$$

For the next step, it is more convenient to write the equations as a list as follows:

$$p_0 = e^{-\lambda\tau}(p_0 + p_1)$$
$$p_1 = e^{-\lambda\tau}[(p_0 + p_1)\lambda\tau + p_2]$$
$$p_2 = e^{-\lambda\tau}\left[(p_0 + p_1)\frac{(\lambda\tau)^2}{2!} + p_2(\lambda\tau) + p_3\right]$$

$$\vdots = \vdots$$

$$p_n = e^{-\lambda\tau}\left[(p_0 + p_1)\frac{(\lambda\tau)^n}{n!} + p_2\frac{(\lambda\tau)^{n-1}}{(n-1)!} + \cdots + p_n(\lambda\tau) + p_{n+1}\right]$$

$$\vdots = \vdots$$

We now construct the probability generating function for the sequence $\{p_n\}$ by multiplying p_n by s^n in the list above, and summing over n by *columns* to give

$$G(s) = \sum_{n=0}^{\infty} p_n s^n = e^{-\lambda\tau}\left[(p_0 + p_1)e^{\lambda\tau s} + p_2 s e^{\lambda\tau s} + p_3 s^2 e^{\lambda\tau s} + \cdots\right]$$

$$= e^{-\lambda\tau(1-s)}\left[p_0 + p_1 + p_2 s + p_3 s^2 + \cdots\right]$$

$$= e^{-\lambda\tau(1-s)}\left[p_0 + \frac{1}{s}\{G(s) - p_0\}\right].$$

Hence, solving this equation for $G(s)$, we obtain

$$G(s) = \frac{p_0(1-s)}{1 - se^{\lambda\tau(1-s)}}.$$

However, this formula for $G(s)$ still contains an unknown constant p_0. For $G(s)$ to be a *pgf*, we must have $G(1) = 1$. Since both the numerator and denominator tend to zero as $s \to 1$, we must use an expansion to cancel a factor $(1-s)$. By expanding $e^{\lambda\tau(1-s)}$ in powers of $(1-s)$, it follows that

$$G(s) = \frac{p_0(1-s)}{1 - s\left\{1 + \lambda\tau(1-s) + \frac{(\lambda\tau)^2}{2!}(1-s)^2 + \cdots\right\}}$$

$$= \frac{p_0}{1 - s\lambda\tau - \frac{(\lambda\tau)^2}{2!}s(1-s) - \cdots}$$

$$\to \frac{p_0}{1 - \lambda\tau}$$

as $s \to 1$. Hence $p_0 = 1 - \lambda\tau$, and

$$G(s) = \frac{(1 - \lambda\tau)(1-s)}{1 - se^{\lambda\tau(1-s)}}.$$

Alternatively, $G(1)$ can be found by differentiating both sides of

$$(1 - se^{\lambda\tau(1-s)})G(s) = p_0(1 - s)$$

with respect to s and substituting $s = 1$. Individual probabilities can be found by expanding $G(s)$ in powers of s. We shall not discuss convergence here, but simply state that the expansion converges for $0 \leq \lambda\tau < 1$, a result to be expected intuitively.

The expected queue length is given by $G'(1)$ (see Section 1.9), but $G'(s)$ is still singular for $s = 1$. Use the alternative method above but this time differentiate twice with respect to s:

$$(1-se^{\lambda\tau(1-s)})G''(s)+2(\lambda\tau s-1)e^{\lambda\tau(1-s)}G'(s)+[2\lambda\tau-(\lambda\tau)^2 s]e^{\lambda\tau(1-s)}G(s)=0.$$

Putting $s = 1$ and remembering that $G(1) = 1$, we obtain

$$\mu = G'(1) = \frac{\lambda\tau(2 - \lambda\tau)}{2(1 - \lambda\tau)}.$$

This can be expressed in terms of the traffic intensity ρ since $1/\lambda$ is the mean inter-arrival time and τ is the actual service time. Thus, $\rho = \lambda\tau$, and the mean queue length is

$$\mu = \frac{\rho(1 - \frac{1}{2}\rho)}{1 - \rho}.$$

7.6 Classification of queues

There are three main factors that characterize queues: the probability distributions controlling arrivals and service times, the number of servers, and queue discipline – first come, first served. If the inter-arrival and service distributions are denoted by G_1 and G_2, and there are n servers then this queue is described as a $G_1/G_2/n$ queue, that is,

arrival distribution, G_1/ service distribution, G_2 / number of servers, n

queue.

If the inter-arrival and service distributions are negative exponentials with densities $\lambda e^{-\lambda t}$ and $\mu e^{-\mu t}$ (that is, both are Poisson processes), then both processes are Markov with parameters λ and μ respectively. The Markov property means that the probability of the next arrival or the probability of service being completed are independent of any previous occurrences. We denote the processes by $M(\lambda)$ and $M(\mu)$ (M for Markov). Hence, the corresponding single-server queue is denoted by $M(\lambda)/M(\mu)/1$, and the n-server queue by $M(\lambda)/M(\mu)/n$.

The single-server queue with Markov inter-arrival but fixed service time τ, discussed in the previous section, is denoted by $M(\lambda)/D(\tau)/1$, where D stands for deterministic.

This can be extended to other distributions. If the service time for a single-server queue has a uniform distribution with density function

$$f(t) = \begin{cases} \mu & 0 \le t \le 1/\mu, \\ 0 & \text{elsewhere,} \end{cases}$$

then this would be described as an $M(\lambda)/U(\mu)/1$ queue.

7.7 A general approach to the $M(\lambda)/G/1$ queue

In this queue, the arrivals are given by the usual Poisson process with parameter λ, and the service time has a general distribution denoted by G. We adopt a different approach in which the queue is viewed as a chain in which a step in the chain occurs when an individual completes service and leaves the queue. This enables us to obtain a fairly general result on the expected length of the stationary queue.

Suppose that the server opens for business at time $t = 0$, and that the first customer is served and leaves at time T_1, the second leaves at time T_2, and so on, so that T_n is the time of departure of the nth customer. Let $X(T_n)$ be the random variable, which represents the number of customers left in the queue after the nth customer has departed. Suppose that during the service time of the $(n + 1)$th customer, that is during the time $T_{n+1} - T_n$, R new customers join the queue. Provided there is a customer in the queue waiting, that is $X(T_n) > 0$, then

$$X(T_{n+1}) = X(T_n) + R - 1 \quad \text{for} \quad X(T_n) > 0, \tag{7.15}$$

where $X(T_n)$ is the number remaining in the queue at the beginning of the service time, R is the number of new arrivals, and (-1) is the reduction for the person who has just been served. If, on the other hand, $X(T_n) = 0$, then the server would be immediately free, in which case

$$X(T_{n+1}) = R \quad \text{for} \quad X(T_n) = 0. \tag{7.16}$$

Equations (7.15) and (7.16) can be combined into the difference equation

$$X(T_{n+1}) = X(T_n) + R - 1 + h(X), \tag{7.17}$$

where

$$h(X) = \begin{cases} 0 & \text{if } X(T_n) > 0 \\ 1 & \text{if } X(T_n) = 0. \end{cases} \tag{7.18}$$

We now suppose that a stationary queue length develops over a long period of time. Take the expected values of the random variables in (7.17):

$$\mathbf{E}[X(T_{n+1})] = \mathbf{E}[X(T_n + R - 1 + h(X))] = \mathbf{E}[X(T_n)] + \mathbf{E}(R) - \mathbf{E}(1) + \mathbf{E}[h(X)].$$

Assuming that a stationary process ultimately exists, we can put

$$\lim_{n\to\infty} \mathbf{E}[X(T_{n+1})] = \lim_{n\to\infty} \mathbf{E}[X(T_n)] = \mathbf{E}(X),$$

say. Hence

$$\mathbf{E}[h(X)] = 1 - \mathbf{E}(R). \tag{7.19}$$

The expected value $\mathbf{E}(R)$ requires careful explanation. In any given service time, say, τ until the next service starts, the probability that r customers arrive is, for the assumed Poisson process with parameter λ,

$$p_r(\tau) = \frac{e^{-\lambda\tau}(\lambda\tau)^r}{r!}.$$

Its mean value is

$$\mu(\tau) = \sum_{r=1}^{\infty} r p_r(\tau) = \sum_{r=1}^{\infty} \frac{e^{-\lambda\tau}(\lambda\tau)^r}{(r-1)!} = \lambda\tau. \tag{7.20}$$

This mean is itself a random variable, which is a function of the random variable of the service time τ. Let the probability density function for the service time distribution be $f(\tau)$. Hence, by the formula for the expected value of a function of a random variable (Section 1.6), the expected value of R is given by

$$\mathbf{E}(R) = \int_0^{\infty} \tau f(\tau)\,d\tau = \lambda\mathbf{E}(S) = \rho, \tag{7.21}$$

say, where $\mathbf{E}(S)$ is the expected value of the random variable S of the service times. The expected value ρ, as in previous sections, is the traffic density, and it is assumed that $\rho < 1$. From (7.19), it follows that

$$\mathbf{E}[h(X)] = 1 - \rho. \tag{7.22}$$

Now take the square of (7.17):

$$\begin{aligned}
X(T_{n+1})^2 &= X(T_n)^2 + (R-1)^2 + h(X)^2 + 2(R-1)h(X) \\
&\quad + 2h(X)X(T_n) + 2(R-1)X(T_n) \\
&= X(T_n)^2 + (R-1)^2 + h(X)^2 + 2(R-1)[h(X) + X(T_n)],
\end{aligned}$$

in which it should be observed that $h(X)X(T_n) = 0$ by equation (7.18). The reason for this step is that, as we shall show, the expected value of R^2 is related to the variance of the service distribution. With this in view, take the expected value of the stationary version of this equation so that, with

$$\lim_{n\to\infty} \mathbf{E}[X(T_{n+1})] = \lim_{n\to\infty} \mathbf{E}[X(T_n)] = \mathbf{E}(X),$$

again

$$\mathbf{E}[(R-1)^2] + \mathbf{E}[h(X)^2] + 2\mathbf{E}[Rh(X)] - 2\mathbf{E}[h(X)] + 2\mathbf{E}[(R-1)X] = 0,$$

or

$$\mathbf{E}(R^2) - 2\mathbf{E}(R) + 1 - \mathbf{E}[h(X)] + 2\mathbf{E}(R)\mathbf{E}[h(X)] + 2\mathbf{E}(R)\mathbf{E}(X) - 2\mathbf{E}(X) = 0, \tag{7.23}$$

since $h(X)^2 = h(X)$ by equation (7.18), and X and R are independent.

We shall solve this equation for $\mathbf{E}(X)$, the expected length of the stationary queue, but first we need to find a formula for $\mathbf{E}(R^2)$.

Let r be a value of the random variable R. We require the second moment of the Poisson distribution, namely

$$
\begin{aligned}
u(\tau) = \sum_{r=1}^{\infty} r^2 p_r(\tau) &= \sum_{r=1}^{\infty} \frac{re^{-\lambda\tau}(-\lambda\tau)^r}{(r-1)!} \\
&= \tau e^{-\lambda\tau} \frac{\mathrm{d}}{\mathrm{d}\tau} \left[\sum_{r=1}^{\infty} \frac{(\lambda\tau)^r}{(r-1)!} \right] \\
&= \tau e^{-\lambda\tau} \frac{\mathrm{d}}{\mathrm{d}\tau} (\lambda\tau e^{\lambda\tau}) = \lambda\tau + (\lambda\tau)^2.
\end{aligned}
$$

or directly from the mean and variance of the Poisson distribution. As before, $u(\tau)$ is a function of the service time τ, so that

$$
\begin{aligned}
\mathbf{E}(R^2) = \int_0^\infty u(\tau) f(\tau)\,\mathrm{d}\tau &= \int_0^\infty [\lambda\tau + (\lambda\tau)^2] f(\tau)\,\mathrm{d}\tau, \\
&= \mathbf{E}(R) + \lambda^2 \int_0^\infty \tau^2 f(\tau)\,\mathrm{d}\tau.
\end{aligned}
$$

The integral in the last result is $V(\tau) + [\mathbf{E}(\tau)]^2$. With S the random variable whose values are given by τ, from the general result for variance (Section 1.6)

$$V(S) = \mathbf{E}(S^2) - [\mathbf{E}(S)]^2 = \int_0^\infty \tau^2 f(\tau)\,\mathrm{d}\tau - \left[\int_0^\infty \tau f(\tau)\,\mathrm{d}\tau \right]^2.$$

Hence

$$\mathbf{E}(R^2) = \mathbf{E}(R) + \lambda^2 V(S) + [\mathbf{E}(R)]^2 = \rho + \rho^2 + \lambda^2 V(S). \tag{7.24}$$

Finally, solving (7.23) for $\mathbf{E}(X)$ and using (7.21), (7.22) and (7.24), we find that

$$
\begin{aligned}
\mathbf{E}(X) &= \frac{\mathbf{E}(R^2) - 2\mathbf{E}(R) + 1 - \mathbf{E}[h(X)] + 2\mathbf{E}(R)\mathbf{E}[h(X)]}{2 - 2\mathbf{E}(R)} \\
&= \frac{\mathbf{E}(R^2) - 2\mathbf{E}(R) + 1 - \mathbf{E}[h(X)] + 2\mathbf{E}(R)(1 - \mathbf{E}(R))}{2 - 2\mathbf{E}(R)} \\
&= \frac{2\rho - \rho^2 + \lambda^2 V(S)}{2(1 - \rho)}. \tag{7.25}
\end{aligned}
$$

The expected value $\mathbf{E}(X)$ is the expected length of the queue, including the person being served, and it is interesting to note that it is a function of the variance of the service time. If the service times have a negative exponential distribution, with density $\mu e^{-\mu t}$ (that is the queue is $M(\lambda)/M(\mu)/1$), then its mean and variance are $\mathbf{E}(S) = 1/\mu$ and $\mathbf{V}(S) = 1/\mu^2$ respectively. The traffic density $\rho = \lambda/\mu$ and, from (7.25),

$$\mathbf{E}(X) = \frac{2\rho - \rho^2 + \rho^2}{2(1 - \rho)} = \frac{\rho}{1 - \rho},$$

which is the mean value obtained in Section 7.3(b).

One implication of (7.25) is that two queues with the same Poisson parameter λ for arrivals, and the same traffic intensities (which means equal expected service times) can have different mean queue lengths. It follows from (7.25) that, for a fixed ρ, $\mathbf{E}(X)$ increases linearly with the variance $\mathbf{V}(S)$. Generally, larger variances lead to longer queues.

Example 7.4. The $M(\lambda)/U/1$ queue has a service time with uniform distribution with density

$$f(t) = \begin{cases} \mu & 0 \le t \le 1/\mu \\ 0 & \text{elsewhere.} \end{cases}$$

Find the expected length of the queue.

We apply formula (7.25) for which we need $\mathbf{E}(R)$, $\mathbf{E}(S)$ and $\mathbf{V}(S)$. The expected value of the random variable of the service times is

$$\mathbf{E}(S) = \int_{-\infty}^{\infty} t f(t)\, \mathrm{d}t = \mu \int_{0}^{1/\mu} t\, \mathrm{d}t = \frac{1}{2\mu},$$

and its variance is

$$\mathbf{V}(S) = \mathbf{E}(S^2) - [\mathbf{E}(S)]^2 = \int_{-\infty}^{\infty} t^2 f(t)\, \mathrm{d}t - \frac{1}{4\mu^2}$$

$$= \mu \int_{0}^{1/\mu} t^2\, \mathrm{d}t - \frac{1}{4\mu^2} = \frac{1}{3\mu^2} - \frac{1}{4\mu^2} = \frac{1}{12\mu^2}.$$

Since the arrivals follow a negative exponential distribution with parameter λ, $\mathbf{E}(R) = \lambda$. Hence, from (7.21)

$$\rho = \mathbf{E}(R)/\mathbf{E}(S) = \lambda/(2\mu),$$

which must be less than 1 for a finite queue. From (7.25), the expected length of the queue is given by

$$\mathbf{E}(X) = \frac{2\rho - \rho^2 + \frac{1}{3}\rho^2}{2(1 - \rho)} = \frac{\rho(3 - \rho)}{3(1 - \rho)}.$$

In this particular queue, the service time is capped at time $1/\mu$, so that no service is allowed to take longer than this time. $\qquad\square$

Problems

7.1. In a single-server queue, a Poisson process for arrivals of intensity $\frac{1}{2}\lambda$ and for service and departures of intensity λ are assumed. For the corresponding stationary process, find

    (a) p_n, the probability that there are n persons in the queue,

    (b) the expected length of the queue,

    (c) the probability that there are not more than two persons in the queue, including the person being served in each case.

7.2. Consider a telephone exchange with a very large number of lines available. If n lines are busy the probability that one of them will become free in small time δt is $n\mu\delta t$. The probability of a new call is $\lambda\delta t$ (that is, Poisson), with the assumption that the probability of multiple calls is negligible. Show that $p_n(t)$, the probability that n lines are busy at time t satisfies

$$\frac{dp_0(t)}{dt} = -\lambda p_0(t) + \mu p_1(t),$$

$$\frac{dp_n(t)}{dt} = -(\lambda + n\mu)p_n(t) + \lambda p_{n-1}(t) + (n+1)\mu p_{n+1}(t), \quad (n \geq 1).$$

In the stationary process, show by induction that

$$p_n = \lim_{t \to \infty} p_n(t) = \frac{e^{-\lambda/\mu}}{n!}\left(\frac{\lambda}{\mu}\right)^n.$$

Identify the distribution.

7.3. For a particular queue, when there are n customers in the system, the probability of an arrival in the small time interval δt is $\lambda_n \delta t + o(\delta t)$. The service time parameter μ_n is also a function of n. If p_n denotes the probability that there are n customers in the steady state queue, show by induction that

$$p_n = p_0 \frac{\lambda_0 \lambda_1 \dots \lambda_{n-1}}{\mu_1 \mu_2 \dots \mu_n},$$

and find an expression for p_0.

    If $\lambda_n = 1/(n+1)$ and $\mu_n = \mu$, a constant, find the expected length of the queue.

7.4. In a baulked queue (see Example 7.1) not more than $m \geq 2$ people are allowed to form a queue. If there are m individuals in the queue, then any further arrivals are turned away. If the arrivals form a Poisson process with parameter λ and the service distribution is exponential with parameter μ, show that the expected length of the queue is

$$\frac{\rho - (m+1)\rho^{m+1} + m\rho^{m+2}}{(1-\rho)(1-\rho^{m+1})},$$

where $\rho = \lambda/\mu$. What is the expected length of the queue if $m = 3$ and $\rho = 1$?

7.5. Consider the single-server queue with arrivals occurring as a Poisson distribution with parameter λ, and service times having and exponential distribution with parameter μ. In the stationary process, the probability p_n that there are n individuals in the queue is given by

$$p_n = \left(1 - \frac{\lambda}{\mu}\right)\left(\frac{\lambda}{\mu}\right)^n, \quad (n = 0, 1, 2, \ldots).$$

Find its probability generating function

$$G(s) = \sum_{n=0}^{\infty} p_n s^n.$$

If $\lambda < \mu$, use this function to determine the mean and variance of the queue length.

7.6. A queue is observed to have an average length of 2.8 individuals including the person being served. Assuming the usual negative exponential distributions for both service times and times between arrivals, what is the traffic density, and the variance of the queue length?

7.7. The non-stationary differential-difference equations for a queue with parameters λ and μ are (see equation (7.1))

$$\frac{dp_0(t)}{dt} = \mu p_1(t) - \lambda p_0(t),$$

$$\frac{dp_n(t)}{dt} = \lambda p_{n-1}(t) + \mu p_{n+1}(t) - (\lambda + \mu) p_n(t),$$

where $p_n(t)$ is the probability that the queue has length n at time t. Let the probability generating function of the distribution $\{p_n(t)\}$ be

$$G(s, t) = \sum_{n=0}^{\infty} p_n(t) s^n.$$

Show that $G(s, t)$ satisfies the equation

$$s\frac{\partial G(s, t)}{\partial t} = (s - 1)(\lambda s - \mu)G(s, t) + \mu(s - 1)p_0(t).$$

Unlike the birth and death processes in Chapter 6, this equation contains the unknown probability $p_0(t)$, which complicates its solution. Show that it can be eliminated to leave the following second-order partial differential equation for $G(s, t)$:

$$s(s - 1)\frac{\partial^2 G(s, t)}{\partial t \partial s} - (s - 1)^2(\lambda s - \mu)\frac{\partial G(s, t)}{\partial s}$$
$$- \frac{G(s, t)}{\partial t} - \lambda(s - 1)^2 G(s, t) = 0.$$

This equation can be solved by Laplace transform methods.

7.8. A call centre has r telephones manned at any time, and the traffic density is $\lambda/(r\mu) = 100$. How many telephones should be manned in order that the expected number of callers waiting at any time should not exceed 4? Assume a stationary process with inter-arrival times of calls and service times for all operators exponential with parameters λ and μ respectively.

7.9. Compare the expected lengths of the two queues $M(\lambda)/M(\mu)/1$ and $M(\lambda)/D(1/\mu)/1$. the queues have parameters such that the mean service time for the former equals the fixed service time in the latter. For which queue would you expect the mean queue length to be the shorter?

7.10. A queue is serviced by r servers, with the inter-arrival time distribution for the queue being exponential with parameter λ and the service times distributions for each server being Poisson with parameter μ. If N is the random variable representing the length of the queue *including* those being served, show that its expected value is

$$\mathbf{E}(N) = p_0\left[\sum_{n=1}^{r-1}\frac{\rho^n}{(n - 1)!} + \frac{\rho^r[r^2 + \rho(1 - r)]}{(r - 1)!(r - \rho)^2}\right],$$

where $\rho = \lambda/\mu$, and

$$p_0 = 1 \bigg/ \left[\sum_{n=0}^{r-1}\frac{\rho^n}{n!} + \frac{\rho^r}{(r - \rho)(r - 1)!}\right]$$

(see equation (7.11)).
If $r = 2$, show that

$$\mathbf{E}(N) = \frac{4\rho}{4 - \rho^2}.$$

For what interval of values of ρ is the expected length of the queue less than the number of servers?

7.11. For a queue with two servers, the probability p_n that there are n persons in the queue is given by

$$p_0 = \frac{2 - \rho}{2 + \rho}, \qquad p_1 = \rho p_0,$$

$$p_n = 2\left(\frac{\rho}{2}\right)^n p_0, \qquad (n \geq 2),$$

where $\rho = \lambda/\mu$ (see Section 7.4). If the random variable X is the number of people in the queue, find its probability generating function. Hence find the mean length of the queue, including those being served.

7.12. The queue $M(\lambda)/D(\tau)/1$, which has a fixed service time of duration τ for every customer, has the probability generating function

$$G(s) = \frac{(1-\rho)(1-s)}{1 - se^{\rho(1-s)}},$$

where $\rho = \lambda\tau \; (0 < \rho < 1)$.

(a) Find the probabilities p_0, p_1, p_2.
(b) Find the expected value and variance of the length of the queue.
(c) Customers are allowed a service time τ, which is such that the expected length of the queue is two individuals. Find the value of the traffic density ρ.

7.13. For the baulked queue, which has a maximum length of m, beyond which customers are turned away, the probabilities that there are n individuals in the queue are given by

$$p_n = \frac{\rho^n(1-\rho)}{1-\rho^{m+1}}, \quad (\rho \neq 1), \qquad p_n = \frac{1}{m+1}, \quad (\rho = 1),$$

for $n = 0, 1, 2, \ldots, m$. Show that the probability generating functions are

$$G(s) = \frac{(1-\rho)[1-(\rho s)^{m+1}]}{(1-\rho^{m+1})(1-\rho s)}, \qquad (\rho \neq 1),$$

and

$$G(s) = \frac{1 - s^{m+1}}{(m+1)(1-s)}, \qquad (\rho = 1).$$

Find the expected value of the queue length including the person being served.

7.14. In Section 7.3(b), the expected length of the queue with parameters λ and μ, including the person being served, was shown to be $\rho/(1-\rho)$. What is the expected length of the queue *excluding* the person being served?

7.15. An $M(\lambda)/M(\mu)/1$ queue is observed over a long period of time. Regular sampling indicates that the mean length of the queue, including the person being served, is 3, whilst the mean waiting time to completion of service by any customer arriving is 10 minutes. What is the mean service time?

7.16. An $M(\lambda)/G/1$ queue has a service time with a gamma distribution with parameters n and α where n is a positive integer. Its density function is

$$f(t) = \frac{\alpha^n t^{n-1} e^{-\alpha t}}{(n-1)!}, \qquad t > 0.$$

Show that, assuming $\rho = \lambda n/\alpha$, the expected length of the queue, including the person being served, is given by

$$\frac{\rho[2n - (n-1)\rho]}{2n(1-\rho)}.$$

7.17. The service in a single-server queue must take at least 2 minutes and must not exceed 8 minutes, and the probability that the service is completed at any time between these times is uniformly distributed over the time interval. The average time between arrivals is 4 minutes, and the arrival distribution is assumed to be Poisson. Calculate the expected length of the queue.

7.18. A person arrives at an $M(\lambda)/M(\mu)/1$ queue. If there are two people in the queue (excluding customers being served) the customer goes away and does not return. If there are there are fewer than two queueing then the customer joins the queue. Show that the expected waiting time for the customer to the start of service is

$$(1 + 2\rho)/[\mu(1 + \rho)],$$

where $\rho = \lambda/\mu$.

7.19. A customer waits for service in a bank in which there are four counters with customers at each counter but otherwise no one is queueing. If the service time distribution is negative exponential with parameter μ for each counter, for how long should the queueing customer have to wait?

7.20. A hospital has two operating theatres dedicated for a particular group of operations. Assuming that the queue is stationary, and that the waiting list and operating time can be viewed as an $M(\lambda)/M(\mu)/1$ queue, show that the expected value of the random variable N representing the length of the queue is given by

$$\mathbf{E}(N) = \frac{\rho^3}{4 - \rho^2}, \qquad \rho = \frac{\lambda}{\mu} < 2.$$

It is observed that the average waiting list is very long at 100 individuals. Why will ρ be very close to 2? Put $\rho = 2 - \varepsilon$ where $\varepsilon > 0$ is small. Show that $\varepsilon \approx 0.02$. A third operating theatre is brought into use with the same operating parameter μ. What effect will this new theatre have on the waiting list eventually?

7.21. Consider the $M(\lambda)/M(\mu)/r$ queue which has r servers such that $\rho = \lambda/\mu < r$. Adapting the method for the single-server queue (Section 7.3 (c)), explain why the average service time for $(n - r + 1)$ customers to be served is $(n - r + 1)/(\mu r)$ if $n \geq r$. What is it if $n < r$? If $n \geq r$, find the average value of the waiting time random variable T until service using

$$\mathbf{E}(T) = \sum_{n=r}^{\infty} \frac{n - r + 1}{r\mu} p_n.$$

What is the average waiting time if service is included?

7.22. Consider the $M(\lambda)/M(\mu)/r$ queue. Assuming that $\lambda < r\mu$, what is the probability in the long term that at any instant there is no one queueing excluding those being served?

7.23. Access to a toll road is controlled by r toll booths. Vehicles approaching the toll booths choose one at random: any toll booth is equally likely to be the one chosen irrespective of the number of cars queueing (perhaps an unrealistic situation). The payment time is assumed to be negative exponential with parameter μ, and vehicles are assumed to approach with parameter λ. Show that, viewed as a stationary process, the queue of vehicles at any toll booth is an $M(\lambda/r)/M(\mu)/1$ queue assuming $\lambda/(r\mu) < 1$. Find the expected number of vehicles queueing at any toll booth. How many cars will be queueing over all booths?

One booth is out of action, and vehicles distribute themselves randomly over the remaining booths. Assuming that $\lambda/[(r - 1)\mu] < 1$, how many extra vehicles can be expected to be queueing overall?

7.24. In an $M(\lambda)/M(\mu)/1$ queue, it is decided that the service parameter μ should be adjusted to make, on average, the slack period 5 times the busy period to allow the server some respite. What should μ be in terms of λ?

7.25. In the baulked queue (see Example 7.1) not more than $m \geq 2$ people (including the person being served) are allowed to form a queue. If there are m individuals in the queue, then any further arrivals are turned away. It is assumed that the service distribution is exponential with rate μ. If $\rho \neq 1$, show that the expected length of the busy periods is given by $(1 - \rho^m)/(\mu - \lambda)$.

7.26. The $M(\lambda)/D(\tau)/1$ queue has a fixed service time τ, and from Section 7.5, its probability generating function is

$$G(s) = \frac{(1 - \lambda\tau)(1 - s)}{1 - se^{\lambda\tau(1-s)}}.$$

Show that the expected length of its busy periods is $\tau/(1 - \lambda\tau)$.

7.27. Viewed as a stationary process, the probability that a queue has length n, including those being served, is given by

$$\mu p_1 - \lambda p_0 = 0,$$

$$\lambda p_{n-1} + (n+1)\mu p_{n+1} - (\lambda + n\mu)p_n = 0, \qquad (n = 1, 2, \dots, r-1),$$

$$\lambda p_{r-1} - (\lambda + r\mu)p_r = 0.$$

Describe a queue to which this probability model would apply. Show that

$$p_n = \rho^n \left/ \left[\sum_{m=0}^{r} \frac{\rho^m}{m!} \right], \right.$$

where $\rho = \lambda/\mu$.

8
Reliability and renewal

8.1 Introduction

The expected life until breakdown occurs of domestic equipment such as television sets, video recorders, washing machines, central heating boilers, etc, is important to the consumer. Failure means the temporary loss of the facility and expense. A television set, for example, has a large number of components, and the failure of just one of them may cause the failure of the set. Increasingly, individual components now have high levels of reliability, and the reduction of the likelihood of failure by the inclusion of alternative components or circuits is probably not worth the expense for a television, which will suffer some deterioration with time anyway. In addition, failure of such equipment is unlikely to be life-threatening.

In other contexts reliability is extremely important for safety or security reasons, as in, for example, aircraft or defence installations, where failure could be catastrophic. In such systems, back-up circuits and components and fail-safe systems are necessary, yet there could still be the remote possibility of both main and secondary failure. Generally, questions of cost, safety, performance, reliability and complexity have to be resolved together with risk assessment.

In medical situations, the interest lies in the survivability of the patient, that is, the expected survival time to death or relapse, under perhaps alternative medical treatments. Again factors such as cost and quality of life can be important.

Reliability is the probability of a device giving a specified performance for an intended period. For a system, we might be interested in the probability that it is still operating after time t, the expected time to failure and the failure rate. There are many external factors that could affect these functions, such as manufacturing defects (which often cause early failure), unexpected operating conditions or operator error.

8.2 The reliability function

Let T be a non-negative random variable, which is the time to failure or *failure time* for a component, device or system. Suppose that the *distribution function* of

T is F so that

$$F(t) = \mathbf{P}\{\text{component fails at or before time } t\},$$
$$= \mathbf{P}\{T \le t\}, \quad t \ge 0.$$

The *density* of T is denoted by f, that is,

$$f(t) = \frac{dF(t)}{dt}. \tag{8.1}$$

It is more convenient – and also more convenient for the consumer or customer – to look at the *reliability* or *survivor function*, $R(t)$, rather than the failure time distribution. The function $R(t)$ is the probability that the device is still operating after time t. Thus

$$R(t) = \mathbf{P}\{T > t\} = 1 - F(t). \tag{8.2}$$

From equation (8.1)

$$f(t) = \frac{dF(t)}{dt} = \lim_{\delta t \to 0} \frac{F(t + \delta t) - F(t)}{\delta t},$$
$$= \lim_{\delta t \to 0} \frac{\mathbf{P}\{T \le t + \delta t\} - \mathbf{P}\{T \le t\}}{\delta t},$$
$$= \lim_{\delta t \to 0} \frac{\mathbf{P}\{t < T \le t + \delta t\}}{\delta t}.$$

Hence, for small δt, $f(t)\delta t$ is approximately the probability that failure occurs in the time interval $(t, t + \delta t)$.

The *failure rate function* or *hazard function* $r(t)$ is defined as that function for which $r(t)\delta t$ is the conditional probability that failure occurs in the interval $(t, t + \delta t)$, *given that the device was operating at time t.* As the limit of this conditional probability, the failure rate function is given by

$$r(t) = \lim_{\delta t \to 0} \frac{\mathbf{P}\{t < T \le t + \delta t | T > t\}}{\delta t}. \tag{8.3}$$

There are two important results relating the reliability and failure rate functions, which will be be proved in the following theorem.

Theorem 8.1.
(i) The failure rate function

$$r(t) = \frac{f(t)}{R(t)}. \tag{8.4}$$

(ii) The reliability function

$$R(t) = \exp\left[-\int_0^t r(s)\, ds\right]. \tag{8.5}$$

Proof.

(i) Use the conditional probability definition (Section 1.3):

$$\mathbf{P}\{A|B\} = \mathbf{P}\{A \cap B\}/\mathbf{P}\{B\}.$$

Thus

$$\mathbf{P}\{t < T \le t + \delta t | T > t\} = \frac{\mathbf{P}\{(t < T \le t + \delta t) \cap (T > t)\}}{\mathbf{P}\{T > t\}}$$

$$= \frac{\mathbf{P}\{t < T \le t + \delta t\}}{\mathbf{P}\{T > t\}}.$$

From (8.2) and (8.3)

$$r(t) = \lim_{\delta t \to 0} \frac{\mathbf{P}\{t < T \le t + \delta t\}}{\delta t} \frac{1}{\mathbf{P}\{T > t\}} = \frac{f(t)}{R(t)}.$$

(ii) Using (8.1) and the previous result

$$r(t) = \frac{f(t)}{R(t)} = \frac{\mathrm{d}F(t)}{\mathrm{d}t} \frac{1}{R(t)} = -\frac{\mathrm{d}R(t)}{\mathrm{d}t} \frac{1}{R(t)} = -\frac{\mathrm{d}}{\mathrm{d}t}(\ln R(t)).$$

Integration of this result gives

$$\ln R(t) = - \int_0^t r(s)\, \mathrm{d}s,$$

since $R(0) = 1$ (the device must be operating at start-up). Hence

$$R(t) = \exp\left[- \int_0^t r(s)\, \mathrm{d}s\right].$$

Hence the failure experience of a device may be summarized by the failure rate function $r(t)$. This completes the proof.

Example 8.1. Whilst in use, an office photocopier is observed to have a failure rate function $r(t) = 2\lambda t$ per hour where $\lambda = 0.00028(\text{hours})^{-2}$. What is the probability that the photocopier is still operational after 20 hours? Find also the associated probability density function $f(t)$ for the problem.

The failure rate function increases linearly with time. By Theorem 8.1(ii),

$$R(t) = \exp\left[- \int_0^t r(s)\, \mathrm{d}s\right] = \exp\left[- \int_0^t 2\lambda t\, \mathrm{d}t\right] = \exp(-\lambda t^2).$$

Thus, the probability that the photocopier is still working after 20 hours is

$$R(20) = \mathrm{e}^{-0.00028 \times (20)^2} = 0.89.$$

By Theorem 8.1(i), the associated probability density function is

$$f(t) = r(t)R(t) = 2\lambda t \mathrm{e}^{-\lambda t^2},$$

which is a Weibull distribution (see Section 1.8) with parameters 2 and λ. □

8.3 Exponential distribution and reliability

For the exponential distribution (see Section 1.8), the probability density function is $f(t) = \lambda e^{-\lambda t}$ $(t \geq 0)$. Thus

$$R(t) = 1 - F(t) = 1 - \int_0^t f(s)\,\mathrm{d}s = 1 + [e^{-\lambda s}]_0^t = e^{-\lambda t}, \qquad (8.6)$$

and

$$r(t) = \frac{f(t)}{R(t)} = \frac{\lambda e^{-\lambda t}}{e^{-\lambda t}} = \lambda.$$

This constant failure rate implies that the probability of failure in a time interval δt, given survival at time t, is independent of the time at which this occurs. Thus, the conditional survival probability for time t beyond, say, time s is

$$\mathbf{P}\{T > s + t | T > s\} = \mathbf{P}\{(T > s + t) \cap (T > s)\}$$

$$= \frac{\mathbf{P}\{T > s + t\}}{\mathbf{P}\{T > s\}}$$

$$= \frac{e^{-\lambda(s+t)}}{e^{-\lambda s}} = e^{-\lambda t},$$

which is the no memory property of the exponential distribution. The implication is that such devices do not deteriorate with age, which means that they can be re-used with the same probability of subsequent failure at any stage in their lives.

8.4 Mean time to failure

In terms of the probability density function, the mean time to failure is

$$\mathbf{E}(T) = \int_0^\infty t f(t)\,\mathrm{d}t. \qquad (8.7)$$

The mean can be expressed in terms of the reliability function $R(t)$ as follows. Since from equations (8.1) and (8.2), $f(t) = -\mathrm{d}R/\mathrm{d}t$, then

$$\mathbf{E}(T) = - \int_0^\infty t \frac{\mathrm{d}R(t)}{\mathrm{d}t}\,\mathrm{d}t = -[t R(t)]_0^\infty + \int_0^\infty R(t)\,\mathrm{d}t$$

after integrating by parts. Assuming that $t R(t) \to 0$ as $t \to \infty$ (this is certainly true for the negative exponential distribution since $R(t) = e^{-\lambda t}$, which tends to zero faster than $t \to \infty$), then

$$\mathbf{E}(T) = \int_0^\infty R(t)\,\mathrm{d}t. \qquad (8.8)$$

For the exponential distribution of the previous section, $R(t) = e^{-\lambda t}$ and

$$\mathbf{E}(T) = \int_0^\infty e^{-\lambda t}\,\mathrm{d}t = 1/\lambda. \qquad (8.9)$$

Example 8.2. Find the mean time to failure for the photocopier of Example 8.1. What is the variance of the failure times?

It was shown that the reliability function was $R(t) = e^{-\lambda t^2}$. Hence the mean time to failure is

$$\mathbf{E}(T) = \int_0^\infty e^{-\lambda t^2}\, dt = \tfrac{1}{2}\sqrt{(\pi/\lambda)} = 52.96 \text{ hours,}$$

using the standard formula for the integral (see the Appendix).
  Now

$$\mathbf{E}(T^2) = \int_0^\infty t^2 f(t)\, dt = -\int_0^\infty t^2 \frac{dR(t)}{dt}\, dt$$

$$= -[t^2 R(t)]_0^\infty + \int_0^\infty 2t R(t)\, dt = \int_0^\infty 2t e^{-\lambda t^2}\, dt$$

$$= 1/\lambda,$$

since $t^2 R(t) = t^2 e^{-\lambda t^2} \to 0$ as $t \to \infty$. Hence the variance is given by

$$\mathbf{V}(T) = \mathbf{E}(T^2) - [\mathbf{E}(T)]^2 = \frac{1}{\lambda} - \frac{\pi}{4\lambda} = \frac{4-\pi}{4\lambda} = 766.4 \text{ (hours)}^2.$$

$\square$

8.5   Reliability of series and parallel systems

In many systems, components can be in series configurations as in Figure 8.1(a), or in parallel configurations as in Figure 8.1(b), in complex series/parallel systems as in Figure 8.1(c), or perhaps with bridges as shown in Figure 8.1(d).
  In the series of components in Figure 8.1(a), the system is only operable if all the components are working; if any one fails then the system fails. Suppose that the system has m components $c_1, c_2, \ldots, c_m$ in series, which operate independently giving independent times to failure, with corresponding reliability functions $R_1(t), R_2(t), \ldots, R_m(t)$. (Generally this is not true as any one failure affects other components in the system.) If T is the time to failure then, since all components have to be operating for the system to operate,

$$R(t) = \mathbf{P}\{T > t\} = \mathbf{P}\{T > t \text{ for } c_1\}\mathbf{P}\{T > t \text{ for } c_2\} \ldots \mathbf{P}\{T > t \text{ for } c_m\}$$

$$= R_1(t) R_2(t) \ldots R_m(t) = \prod_{n=1}^m R_n(t). \tag{8.10}$$

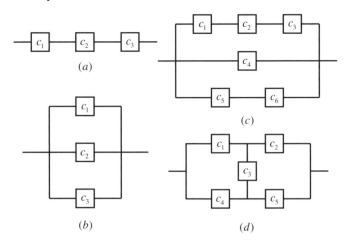

Fig. 8.1 (a) Components in series. (b) Components in parallel. (c) Mixed series/parallel system. (d) Mixed system with bridge.

This is simply the product of the individual reliability functions. The failure rate function is

$$r(t) = -\frac{d}{dt} \ln R(t) = -\frac{d}{dt} \ln \prod_{n=1}^{m} R_n(t) = -\frac{d}{dt} \sum_{n=1}^{m} \ln R_n(t) = \sum_{n=1}^{m} r_n(t),$$

$$(8.11)$$

which is the sum of the individual failure rate functions.

For the parallel configuration in Figure 8.1(b), it is easier to start from the failure distribution function $F(t)$. Since system failure can only occur when *all* components have failed,

$$F(t) = \mathbf{P}\{T \le t\} = \mathbf{P}\{\text{system fails at or before time } t\},$$
$$= \mathbf{P}\{\text{all components fail at or before time } t\},$$
$$= F_1(t)F_2(t) \dots F_m(t) = \prod_{n=1}^{m} F_n(t),$$

where $F_m(t)$ is the failure rate function for c_n. Consequently, the reliability function for m parallel components is

$$R(t) = 1 - F(t) = 1 - \prod_{n=1}^{m} F_n(t) = 1 - \prod_{n=1}^{m} (1 - R_n(t)). \qquad (8.12)$$

A formula for $r(t)$ in terms of the individual functions $r_n(t)$ can be found but it is not particularly illuminating.

Example 8.3. If each component c_m ($m = 1, 2, \ldots, n$) in a series arrangement has an exponentially distributed failure time with parameter λ_m respectively, find the reliability function and the mean time to failure if the failure times are independent.

The reliability function for c_m is $R_m(t) = e^{-\lambda_m t}$. Hence, by equation (8.10), the reliability function for the series of components is

$$R(t) = \prod_{m=1}^{n} R_m(t) = e^{-\lambda_1 t} e^{-\lambda_2 t} \cdots e^{-\lambda_n t} = e^{-(\lambda_1 + \lambda_2 + \cdots + \lambda_n)t},$$

which is the same reliability function for that of a single component with exponentially distributed failure time with parameter $\sum_{m=1}^{n} \lambda_m$. Consequently, by equation (8.9), the mean time to failure is $1 / \sum_{m=1}^{n} \lambda_m$. □

Example 8.4. Three components in parallel, c_1, c_2, c_3, have times to failure that are independent and have identical exponential distributions with parameter λ. Find the mean time to failure.

From (8.12) the reliability function is, since $R_n(t) = e^{-\lambda t}$ ($n = 1, 2, 3$),

$$R(t) = 1 - (1 - e^{-\lambda t})^3.$$

The mean time to failure is

$$\mathbf{E}(T) = \int_0^{\infty} R(t)\, dt = = \int_0^{\infty} \left[1 - (1 - e^{-\lambda t})^3 \right] dt,$$

$$= \int_0^{\infty} \left[1 - (1 - 3e^{-\lambda t} + 3e^{-2\lambda t} - e^{-3\lambda t}) \right] dt,$$

$$= \int_0^{\infty} \left[3e^{-\lambda t} - 3e^{-2\lambda t} + e^{-3\lambda t} \right] dt$$

$$= \left[\frac{3}{\lambda} - \frac{3}{2\lambda} + \frac{1}{3\lambda} \right] = \frac{11}{6\lambda}$$

□

For the mixed system with independent components as shown in Figure 8.1(c), each series subsystem has its own reliability function. Thus, the series c_1, c_2, c_3 has the reliability function $R_1(t) R_2(t) R_3(t)$ and c_5, c_6 the reliability function $R_5(t) R_6(t)$. These composite components are then in parallel with c_4. Hence, by (8.12), the reliablility function for the system is

$$R(t) = 1 - (1 - R_1(t) R_2(t) R_3(t))(1 - R_4(t))(1 - R_5(t) R_6(t)).$$

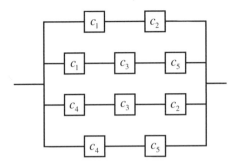

Fig. 8.2 *Series/parallel system equivalent to the bridged system in Figure 8.1(d).*

In Figure 8.1(d) the effect of the bridge containing c_3 is equivalent to four composite components in parallel from the failure potential point of view. We can re-draw the equivalent series/parallel system without the bridge as shown in Figure 8.2. It can be shown that the reliability function is

$$R(t) = 1 - (1 - R_1(t)R_2(t))(1 - R_1(t)R_3(t)R_5(t)) \cdot$$
$$\cdot (1 - R_4(t)R_3(t)R_2(t))(1 - R_4(t)R_5(t)).$$

8.6 Renewal processes

Consider a system that has a single component which, when it fails, is replaced by an identical component with the same lifetime probability distribution. It is assumed that there is no delay at replacement. In this *renewal process* we might ask: how many renewals do we expect in a given time t? Let $N(t)$ be the number of renewals up to and including time t (it is assumed that the first component is installed at start-up at time $t = 0$). Let T_r be a non-negative random variable that represents the time to failure of the rth component measured from its introduction. The time to failure of the first n components, S_n, is given by

$$S_n = T_1 + T_2 + \cdots + T_n.$$

Assuming that the random variables are independent, each with distribution function $F(t_i)$, $(i = 1, 2, \ldots n)$, then the joint distribution function is the probability that $T_1 \leq t_1, T_2 \leq t_2, \ldots, T_n \leq t_n$, that is,

$$\mathbf{P}\{T_1 \leq t_1, T_2 \leq t_2, \ldots, T_n \leq t_n\} = F(t_1)F(t_2) \cdots F(t_n),$$

since $T_1, T_2, \ldots, T_n$ are independent random variables. The corresponding probability density function is $f(t_1)f(t_2) \cdots f(t_n)$ where $f(t_i) = \mathrm{d}F(t_i)/\mathrm{d}t$ $(i = 1, 2, \ldots n)$.

We want to construct a formula for $P\{S_n \leq t\}$. Consider first the probability

$$P\{T_1 + T_2 \leq t\} = \int\int_{\substack{x \geq 0, y \geq 0 \\ x+y \leq t}} f(x)f(y)\,dx\,dy,$$

$$= \int_0^t \int_0^{t-y} f(x)f(y)\,dx\,dy,$$

reversing the order of integration. Hence

$$P\{T_1 + T_2 \leq t\} = \int_0^t F(t-y)f(y)\,dy,$$

where the joint density function is integrated over the triangle shown in Figure 8.3. If we put

$$P\{T_1 + T_2 \leq t\} = F_2(t),$$

then

$$F_2(t) = \int_0^t F_1(t-y)f(y)\,dy, \tag{8.13}$$

where $F_1(t) = F(t)$. We can now repeat this process to obtain the next distribution

$$F_3(t) = P\{S_3 \leq t\} = \int_0^t F_2(t-y)f(y)\,dy,$$

and, in general,

$$F_n(t) = P\{S_n \leq t\} = \int_0^t F_{n-1}(t-y)f(y)\,dy. \tag{8.14}$$

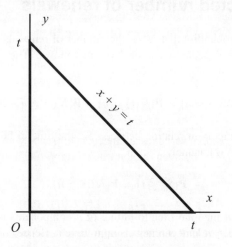

Fig. 8.3 *Area of integration for non-negative random variables.*

Example 8.5. The lifetimes of components in a renewal process with instant renewal are identically and independently distributed with constant failure rate λ. Find the probability that at least two components have been replaced at time t.

Since the components have constant failure rates, then they must have exponentially distributed failure times with parameter λ. Hence $f(t) = \lambda e^{-\lambda t}$ and $F(t) = 1 - e^{-\lambda t}$. Two components or more will have failed if $S_2 = T_1 + T_2 \le t$. Hence by (8.15)

$$P\{S_2 \le t\} = F_2(t) = \int_0^t F(t - y) f(y) \, dy,$$

$$= \int_0^t (1 - e^{-\lambda(t-y)}) \lambda e^{-\lambda y} \, dy,$$

$$= \lambda \int_0^t (e^{-\lambda y} - e^{-\lambda t}) \, dy,$$

$$= \lambda \left[-\frac{e^{-\lambda y}}{\lambda} - y e^{-\lambda t} \right]_0^t,$$

$$= 1 - e^{-\lambda t} - \lambda t e^{-\lambda t} = 1 - (1 + \lambda t) e^{-\lambda t}. \quad (8.15)$$

□

The general formula for $F_n(t)$ is derived in the next section.

8.7 Expected number of renewals

We can find the probability that $N(t)$, the number of renewals up to and including t, is n, and the expected value of $N(t)$, as follows. The probability can be expressed as the following difference:

$$\mathbf{P}\{N(t) = n\} = \mathbf{P}\{N(t) \ge n\} - \mathbf{P}\{N(t) \ge n + 1\}.$$

We now use the important relation between S_n, the time to failure of the first n components, and $N(t)$, namely

$$\mathbf{P}\{S_n \le t\} = \mathbf{P}\{N(t) \ge n\}.$$

This is true since if the total time to failure of n components is less than t, then there must have been at least n renewals up to time t. Hence

$$\mathbf{P}\{N(t) = n\} = \mathbf{P}\{S_n \le t\} - \mathbf{P}\{S_{n+1} \le t\} = F_n(t) - F_{n+1}(t),$$

by equation (8.14). Thus, the expected number of renewals by time t is

$$\mathbf{E}(N(t)) = \sum_{n=1}^{\infty} n\left[F_n(t) - F_{n+1}(t)\right],$$

$$= [F_1(t) + 2F_2(t) + 3F_3(t) + \cdots] - [F_2(t) + 2F_3(t) + \cdots],$$

$$= F_1(t) + F_2(t) + F_3(t) + \cdots,$$

$$= \sum_{n=1}^{\infty} F_n(t), \tag{8.16}$$

which is simply the sum of the individual distribution functions that are given iteratively by (8.14). This is a general formula for any cumulative distribution function $F_n(t)$. The corresponding sequence of density functions is given by

$$f_n(t) = \frac{\mathrm{d}F_n(t)}{\mathrm{d}t} = \frac{\mathrm{d}}{\mathrm{d}t}\int_0^t F_{n-1}(t-y)f(y)\,\mathrm{d}y,$$

$$= F_n(0)f(t) + \int_0^t F'_{n-1}(t-y)f(y)\,\mathrm{d}y,$$

$$= \int_0^t f_{n-1}(t-y)f(y)\,\mathrm{d}y, \tag{8.17}$$

since $F_n(0) = 0$. Note that this is the same iterative formula as for the probability distibutions given by equation (8.14).

Suppose now that the failure times are exponentially distributed with parameter λ: this is essentially a Poisson process. From equation (8.17),

$$f_2(t) = \frac{\mathrm{d}F_2(t)}{\mathrm{d}t} = \lambda^2 t e^{-\lambda t}. \tag{8.18}$$

This suggests that the density function $f_n(t)$ is a gamma distribution with parameters (n, λ). We can prove this by induction. Assume that

$$f_n(t) = \frac{\lambda^n}{(n-1)!}t^{n-1}e^{-\lambda t}.$$

Then, from (8.19),

$$\int_0^t \frac{\lambda^n}{(n-1)!}(t-y)^{n-1}e^{-\lambda(t-y)}\lambda e^{-\lambda y}\,\mathrm{d}y = \frac{\lambda^{n+1}}{(n-1)!}e^{-\lambda t}\int_0^t (t-y)^{n-1}\,\mathrm{d}y,$$

$$= \frac{\lambda^{n+1}}{n!}(t-y)^n e^{-\lambda t} = f_{n+1}(t).$$

Since the result has already been established for $f_2(t)$ in equation (8.18), the result is true by induction for $n = 3, 4, \ldots$. Thus

$$\sum_{n=1}^{\infty} f_n(t) = \sum_{n=1}^{\infty} \frac{\lambda^n}{(n-1)!}t^{n-1}e^{-\lambda t} = \lambda.$$

Finally, from (8.19)

$$E(N(t)) = \sum_{n=1}^{\infty} F_n(t) = \sum_{n=1}^{\infty} \int_0^t f_n(t)\, dt,$$

$$= \int_0^t \sum_{n=1}^{\infty} f_n(t)\, dt = \int_0^t \lambda\, dt = \lambda t,$$

a result that could have been anticipated for a Poisson process.

The expected value of $N(t)$ is known as the *renewal function* for the process.

Problems

8.1. The lifetime of a component has a uniform density function given by

$$f(t) = \begin{cases} 1/(t_1 - t_0) & \text{if } 0 < t_0 < t < t_1 \\ 0 & \text{otherwise} \end{cases}$$

For all $t > 0$, obtain the reliability function $R(t)$ and the failure rate function $r(t)$ for the component. Obtain the expected life of the component.

8.2. Find the reliability function $R(t)$ and the failure rate function $r(t)$ for the gamma density

$$f(t) = \lambda^2 t e^{-\lambda t}, \qquad t > 0.$$

How does $r(t)$ behave for large t? Find the mean and variance of the time to failure.

8.3. A failure rate function is given by

$$r(t) = \frac{ct}{a^2 + t^2}, \qquad t \geq 0, \quad a > 0, \quad c > 1,$$

where a and c are parameters. The rate of failures peaks at $t = a$ and then declines towards zero as $t \to \infty$: failure becomes less likely with time (see Figure 8.4). Find the reliability function, and the corresponding probability density.

8.4. A piece of office equipment has a piecewise failure rate function given by

$$r(t) = \begin{cases} 2\lambda_1 t, & 0 < t \leq t_0, \\ 2(\lambda_1 - \lambda_2)t_0 + 2\lambda_2 t, & t > t_0 \end{cases} \qquad \lambda_1, \lambda_2 > 0.$$

Find its reliability function.

8.5. A laser printer is observed to have a failure rate function $r(t) = 2\lambda t$ per hour whilst in use, where $\lambda = 0.00021 (\text{hours})^{-2}$. What is the probability that the printer is working after 40 hours of use? Find the probability density function for the time to failure. What is the expected time before the printer will need maintenance?

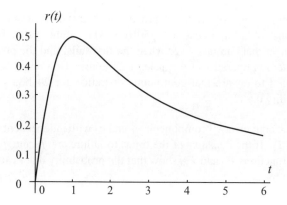

Fig. 8.4 *Failure rate distribution $r(t)$ with $a = 1$ and $c = 1$.*

8.6. The density for a reliability process is assumed to be gamma with parameters λ and n, that is,

$$f(t) = \frac{\lambda(\lambda t)^{n-1}e^{-\lambda t}}{(n-1)!}, \qquad t > 0, \quad \lambda > 0, n > 0.$$

Show that the reliability function is given by

$$R(t) = e^{-\lambda t}\sum_{r=0}^{n-1}\frac{\lambda^r t^r}{r!}.$$

Find the failure rate function and show that $\lim_{t\to\infty} r(t) = \lambda$. What is the expected time to failure?

8.7. A generator has an exponentially distributed failure time with parameter λ_f and the subsequent repair time is exponentially distributed with parameter λ_r. The generator is started up at time $t = 0$. What is the mean time for the generator to fail and the mean time from $t = 0$ for it to be operational again?

   A stand-by generator with exponentially distributed failure time with parameter λ_s is immediately switched on to provide continuous power when the main generator fails. Find the probability that at time τ, measured from the breakdown of the main generator, the stand-by generator breaks down before the main one is repaired.

8.8. A hospital takes a grid supply of electricity, which has a constant failure rate λ. This supply is backed up by a stand-by generator, which has a gamma distributed failure time with parameters μ and 2. Find the reliability function $R(t)$ for the whole electricity supply. Assuming that time is measured in hours, what should the relation between the parameters λ and μ be in order that $R(1000) = 0.999$?

8.9. The components in a renewal process with instant renewal are identical with constant failure rate $\lambda = (1/50)(\text{hours})^{-1}$. If the system has one spare component that can take over when the first fails, find the probability that the system is operational for at least 24 hours. How many spares should be carried to ensure that continuous operation for 24 hours occurs with probability 0.98?

8.10. A device contains two components c_1 and c_2 with independent failure times T_1 and T_2. If the densities of the times to failure are f_1 and f_2 with distribution functions F_1 and F_2, show that the probability that c_1 fails before c_2 is given by

$$P\{T_1 < T_2\} = \int_{y=0}^{\infty} \int_{x=0}^{y} f_1(x) f_2(y) \, dx \, dy,$$
$$= \int_{y=0}^{\infty} F_1(y) f_2(y) \, dy,$$
$$= \frac{\lambda_1}{\lambda_1 + \lambda_2}.$$

Find the probability $P\{T_1 < T_2\}$ in the cases:

(a) both failure times are exponentially distributed with parameters λ_1 and λ_2;

(b) both failure times have gamma distributions with parameters $(\lambda_1, 2)$ and $(\lambda_2, 2)$.

8.11. Let T be a random variable for the failure time of a component. Suppose that the distribution function of T is

$$F(t) = \mathbf{P}(T \le t), \quad t \ge 0,$$

with density

$$f(t) = \alpha_1 e^{-\lambda_1 t} + \alpha_2 e^{-\lambda_2 t}, \qquad \alpha_1, \alpha_2 > 0, \qquad \lambda_1, \lambda_2 > 0,$$

where the parameters satisfy

$$\frac{\alpha_1}{\lambda_1} + \frac{\alpha_2}{\lambda_2} = 1.$$

Find the reliability function $R(t)$ and the failure rate function $r(t)$ for this 'double' exponential distribution. How does $r(t)$ behave as $t \to \infty$?

8.12. The lifetimes of components in a renewal process with instant renewal are identically distributed with constant failure rate λ. Find the probability that at least three components have been replaced at time t.

8.13. The lifetimes of components in a renewal process with instant renewal are identically distributed with a failure rate that has a uniform distribution with density

$$f(t) = \begin{cases} \frac{1}{k} & 0 < t < k \\ 0 & \text{elsewhere} \end{cases}$$

Find the probability that at least two components have been replaced at time t.

8.14. The lifetimes of components in a renewal process with instant renewal are identically distributed each with reliability function

$$R(t) = \frac{1}{2}(e^{-\lambda t} + e^{-2\lambda t}), \qquad t \geq 0, \qquad \lambda > 0.$$

Find the probability that at least two components have been replaced by time t.

8.15. The lifetime of a component in a renewal process has a failure rate with a Weibull density

$$f(t) = \alpha\beta t^{\beta-1}e^{-\alpha t^{\beta}} \qquad (t \geq 0, \alpha, \beta \geq 0).$$

Find the reliability and failure rate functions $R(t)$ and $r(t)$.

9

Branching and other random processes

9.1 Introduction

Branching processes are concerned with the generational growth and decay of populations. The populations could be mutant genes, neutrons in a nuclear chain reaction or birds or animals that have annual cycles of births. As the name implies a branching process creates a *tree* with branches that can split into other branches at each step or at each generation in a chain. In this chapter we shall look at a simplified problem in which the process starts from a single individual that generates the tree. Such a process models a discrete application, such as cellular growth, rather than an animal population in which both births and deaths are taking place continuously in time.

9.2 Generational growth

We assume that the single individual has known probabilities of producing a given number of descendants at a given time, and produces no further descendants. In turn, these descendants each produce further descendants at the next subsequent time with the same probabilities. The process carries on in the same way, creating successive *generations* as indicated in Figure 9.1. At each step, there is a probability p_j that any individual creates j descendants, and it is assumed that this probability distribution $\{p_j\}$ is the same for every individual at every generation. We are interested in the population size in generation n: the earlier generations have either died out or are not counted in the process.

Let X_n be a random variable representing the population size of generation n. Since the process starts with one individual, then $X_0 = 1$. In the representation of a particular process in Figure 9.1,

$$X_1 = 3, \qquad X_2 = 7, \qquad X_3 = 12, \ldots.$$

As described, the generations evolve in step, and occur at the same time steps.

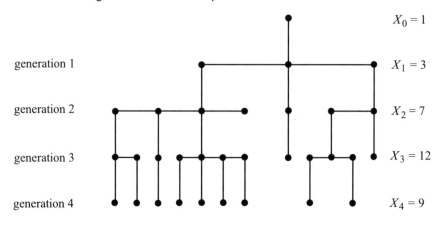

Fig. 9.1 *Generational growth in a branching process.*

Suppose that the probability distribution of descendant numbers $\{p_j\}$ has the probability generating function

$$G(s) = \sum_{j=0}^{\infty} p_j s^j. \tag{9.1}$$

($G(s)$ is really $G_1(s)$ but the suffix is usually suppressed.) Since $X_0 = 1$, $G(s)$ is the generating function for the random variable X_1. For the second generation it is assumed that each descendant has the same probability p_j of independently creating j descendants. Let $G_2(s)$ be the generating function of the random variable X_2, which is the sum of X_1 random variables (the descendants of X_0) which, in turn, are denoted by the independent random variables $Y_1, Y_2, \ldots, Y_{X_1}$, so that

$$X_2 = Y_1 + Y_2 + \cdots + Y_{X_1}.$$

Thus

$$P(Y_k = j) = p_j, \qquad P(X_1 = m) = p_m.$$

Let

$$P(X_2 = n) = h_n.$$

Then, using the partition law (Section 1.3),

$$h_n = P(X_2 = n) = \sum_{r=0}^{\infty} P(X_2 = n | X_1 = r) P(X_1 = r)$$

$$= \sum_{r=0}^{\infty} p_r P(X_2 = n | X_1 = r).$$

Now multiply h_n by s^n and sum over n to create the generating function $G_2(s)$:

$$G_2(s) = \sum_{n=0}^{\infty} h_n s^n = \sum_{r=0}^{\infty} p_r \sum_{n=0}^{\infty} P(X_2 = n | X_1 = r) s^n \qquad (9.2)$$

Consider the latter summation in which r is a fixed integer. Then, by Section 1.9 where the probability generating function is defined,

$$\sum_{n=0}^{\infty} P(X_2 = n | X_1 = r) s^n = E(s^{X_2}) = E(s^{Y_1 + Y_2 + \cdots + Y_r})$$

$$= E(s^{Y_1}) E(s^{Y_2}) \dots E(s^{Y_r}) \quad \text{(since } Y_i \text{ idd)}$$

$$= G(s) G(s) \dots G(s) \quad \text{(multiplied } r \text{ times)}$$

$$= [G(s)]^r.$$

This result also follows from the observation that the distribution of X_2 is the *convolution* of the distribution $\{p_j\}$ with itself r times. Finally, equation (9.2) becomes

$$G_2(s) = \sum_{r=0}^{\infty} p_r [G(s)]^r = G(G(s)). \qquad (9.3)$$

This result holds between any two successive generations. Thus, if $G_m(s)$ is the probability generating function of X_m, then

$$G_m(s) = G_{m-1}(G(s)) = G(G(\dots (G(s)) \dots)), \qquad (9.4)$$

which has $G(s)$ nested m times on the right-hand side. This general result will not be proved here. In this type of branching process, $G(s)$ is the probability generating function of the probability distribution of the numbers of descendants from any individual in any generation to the next generation.

Example 9.1. Suppose in a branching process that any individual has a prob-ability $p_j = 1/2^{j+1}$ $(j = 0, 1, 2, \dots)$ of producing j descendants in the next generation. Find the generating function for X_n, the random variable represent-ing the number of descendants in the nth generation given that $X_0 = 1$.

From (9.1), the probability generating function for X_1 is

$$G(s) = \sum_{j=0}^{\infty} p_j s^j = \sum_{j=0}^{\infty} \frac{1}{2^{j+1}} s^j = \frac{1}{2 - s},$$

after summing the geometric series. From (9.3)

$$G_2(s) = \frac{1}{2 - G(s)} = \frac{1}{2 - \frac{1}{2-s}} = \frac{1}{2 - \frac{1}{2-s}} = \frac{2 - s}{3 - 2s},$$

$$G_3(s) = \frac{2 - G(s)}{3 - 2G(s)} = \frac{3 - 2s}{4 - 3s}, \qquad G_4(s) = \frac{4 - 3s}{5 - 4s}.$$

Generally

$$G_n(s) = G(G_{n-1}(s)) = \frac{1}{2 - G_{n-1}(s)} = \frac{1}{2 - \frac{1}{2 - G_{n-2}(s)}} = \cdots,$$

which is an example of a *continued fraction*.

Looking at the expressions for $G_2(s)$, $G_3(s)$ and $G_4(s)$, we might speculate that

$$G_n(s) = \frac{n - (n - 1)s}{n + 1 - ns}.$$

This result can be established by an induction argument outlined as follows. Assume that the formula holds for n. Then

$$\begin{aligned}
G(G_n(s)) &= \frac{1}{2 - [n - (n - 1)s]/[n + 1 - ns]} \\
&= \frac{n + 1 - ns}{2[n + 1 - ns] - [n - (n - 1)s]} \\
&= \frac{n + 1 - ns}{n + 2 - (n + 1)s},
\end{aligned}$$

which is the formula for $G_{n+1}(s)$; that is, n is replaced by $n + 1$ in the expression for $G_n(s)$. Hence, by induction on the integers, the result is true since we have confirmed it directly for $n = 2$ and $n = 3$.

The power series expansion for $G_n(s)$ can be found by using the binomial theorem. Thus

$$\begin{aligned}
G_n(s) &= \frac{n - (n - 1)s}{n + 1 - ns} = \frac{n - (n - 1)s}{n + 1} \left[1 - \frac{ns}{n + 1} \right]^{-1} \\
&= \frac{1}{n + 1}[n - (n - 1)s] \sum_{r=0}^{\infty} \left(\frac{ns}{n + 1} \right)^r \\
&= \sum_{r=0}^{\infty} \left(\frac{n}{n + 1} \right) s^r - \sum_{r=0}^{\infty} \left(\frac{n - 1}{n + 1} \right) \left(\frac{n}{n + 1} \right)^r s^{r+1} \\
&= \frac{n}{n + 1} + \sum_{r=1}^{\infty} \frac{n^{r-1}}{(n + 1)^{r+1}} s^r.
\end{aligned}$$

The probability that the population of generation n has size r is the coefficient of s^r, which is $n^{r-1}/(n + 1)^{r+1}$ for $r \geq 1$. The probability of extinction for generation n is $G_n(0) = n/(n + 1)$, and this approaches 1 as $n \to \infty$, which means that the probability of ultimate extinction is certain for this branching process. $\square$

9.3 Mean and variance

The mean and variance of the population size of the nth generation can be obtained as fairly simple general formulae, as functions of the mean and variance of

X_1, the random variable of the population size of the first generation. These are μ and σ^2, say, which in terms of the probability generating function $G(s)$ are (see Section 1.9)

$$\mu = \mu_1 = \mathbf{E}(X_1) = G'(1),\tag{9.5}$$

$$\sigma^2 = \sigma_1^2 = \mathbf{V}(X_1) = G''(1) + G'(1) - [G'(1)]^2 = G''(1) + \mu - \mu^2.\tag{9.6}$$

From the previous section, the probability generating function of X_2 is $G(G(s))$. Hence, if μ_n is the mean of X_n, then

$$\mu_2 = \mathbf{E}(X_2) = \frac{d}{ds}G_2(s)|_{s=1} = \frac{d}{ds}[G(G(s))]_{s=1} = G'(1)G'(1) = [G'(1)]^2 = \mu^2,$$

using the chain rule in calculus, and noting that $G(1) = 1$. The method can be repeated for $\mathbf{E}(X_3)$, $\mathbf{E}(X_4)$, …. Thus, μ_n, the mean population size of the nth generation is given by

$$\mu_n = \mathbf{E}(X_n) = \frac{d}{ds}G_n(s)|_{s=1} = \frac{d}{ds}G(G_{n-1}(s))|_{s=1} = G'(1)G'_{n-1}(1)$$

$$= \mu\mu_{n-1} = \mu^2\mu_{n-2} = \cdots = \mu^n\tag{9.7}$$

since $G_{n-1}(1) = 1$.

The variance of the population size of the nth generation is σ_n^2, say, where

$$\sigma_n^2 = \mathbf{V}(X_n) = G_n''(1) + G_n'(1) - [G_n'(1)]^2 = G_n''(1) + \mu^n - \mu^{2n}.\tag{9.8}$$

We can obtain a formula for σ_n^2 as follows. Differentiate

$$G_n'(s) = G'(G_{n-1}(s))G_{n-1}'(s)$$

again, to obtain

$$G_n''(1) = G''(1)[G_{n-1}'(1)]^2 + G'(1)G_{n-1}''(1)$$

$$= (\sigma^2 - \mu + \mu^2)\mu^{2n-2} + \mu G_{n-1}''(1)\tag{9.9}$$

using (9.6). Equation (9.9) is a first-order linear difference equation for $G_n''(1)$.

Write the equation as

$$G_n''(1) - \mu G_{n-1}''(1) = k\mu^{2n-2}, \quad k = \sigma^2 - \mu + \mu^2.\tag{9.10}$$

There are two cases to consider separately: $\mu = 1$ and $\mu \neq 1$.

(i) $\mu \neq 1$. The corresponding homogeneous equation to (9.10) has the solution $G_n''(1) = B\mu^n$, where B is a constant. For the particular solution find the constant C such that $G_n''(1) = C\mu^{2n}$ satisfies the equation. It is easy to show that

$$C = \frac{k}{\mu(\mu - 1)}.$$

Hence

$$G_n''(1) = B\mu^n + \frac{(\sigma^2 - \mu + \mu^2)\mu^{2n}}{\mu(\mu - 1)}.$$

Since $G''(1) = \sigma^2 - \mu + \mu^2$, it follows that

$$B = -\frac{\sigma^2 - \mu + \mu^2}{\mu(\mu - 1)},$$

so that

$$G_n''(1) = \frac{\mu^n(\sigma^2 - \mu + \mu^2)(\mu^n - 1)}{\mu(\mu - 1)}, \qquad (9.11)$$

where it is now obvious why $\mu \neq 1$ in this formula. Finally, from (9.8) and (9.11),

$$\sigma_n^2 = G_n''(1) + \mu^n - \mu^{2n} = \frac{\sigma^2 \mu^{n-1}(\mu^n - 1)}{\mu - 1}.$$

(ii) $\mu = 1$. The equation becomes

$$G_n''(1) - G_n''(1) = \sigma^2.$$

The general solution is now

$$G_n''(1) = A + \sigma^2 n.$$

Since $G''(1) = \sigma^2$, it follows that $A = 0$. Hence, from (9.8),

$$\sigma_n^2 = G_n''(1) = n\sigma^2. \qquad (9.12)$$

Example 9.2. For the branching process with probability $p_j = 1/2^{j+1}$, $j = 0, 1, 2, \ldots$ that any individual has j descendants in the next generation (see Example 9.1), find the mean and variance of the population size of the nth generation.

From Example 9.1, the probability generating function of X_1 is $G(s) = 1/(2 - s)$. Its mean is $\mu = G'(1) = 1$. Hence, by equation (9.7), the mean value of the size of the population of the nth generation is $\mu_n = \mu^n = 1$.

The variance of X_1 is given by (see equation (9.6))

$$\sigma^2 = G''(1) + \mu - \mu^2 = 2.$$

Hence, by equation (9.12), $\sigma_n^2 = 2n$.

Note that, although the probability of ultimate extinction is certain for this branching process, the mean remains 1 for all n, although the variance indicates an increasing spread of values as n increases. □

9.4 Probability of extinction

In Example 9.1, the probability of ultimate extinction for a branching process with the probabilities $p_j = 1/2^{j+1}$ was shown to be certain for this particular process. A general result concerning possible extinction of a population can be found if the probability generating function $G(s)$ for X_1, the random variable of the population size of the first generation, is known.

The probability of extinction by generation n is $G_n(0)$ where $G_n(s)$ is the probability generating function of this generation. $G_n(0)$ must be a non-decreasing function of n since this probability includes the probability that extinction has occurred at an earlier generation. The sequence $\{G_n(0)\}$ is bounded and monotonic increasing with n, and, by a result in analysis, must tend to a limit g, say, as $n \to \infty$.

By equation (9.40), $G_n(0) = G(G_{n-1}(0))$. If $n \to \infty$ on both sides of this equation so that $G_n(0) \to g$ as $n \to \infty$, then g satisfies

$$g = G(g). \tag{9.13}$$

This equation always has the solution $g = 1$, since $G(1) = 1$, but it may have other solutions for $0 \le g < 1$. If it does, which solution should be chosen? Let g_0 be any non-negative solution of equation (9.13). Now any probability generating function is an increasing function of s for $0 \le s \le 1$, since every coefficient of the power series is a probability. Thus

$$G(0) < G(g_0) = g_0,$$
$$G_2(0) = G(G(0)) < G(g_0) = g_0,$$
$$\cdots$$
$$G_n(0) = G(G_{n-1}(0)) < G(g_0) = g_0,$$
$$\cdots \quad .$$

In the limit, as $n \to \infty$, we must have $g \le g_0$ for every solution g_0. We conclude that g must be the *smallest* solution of $g = G(g)$.

Graphically, the smallest positive value can be found by drawing the line $y = x$ and the curve $y = G(x)$ and finding the intersection point that has the smallest positive x.

Example 9.3. For a certain branching process with $X_0 = 1$, the probability generating function for X_1 is given by

$$G(s) = \frac{1}{(2 - s)^2}.$$

Find the probability that the population ultimately becomes extinct. Find also the mean and variance of the populations in the nth generation. Interpret the results.

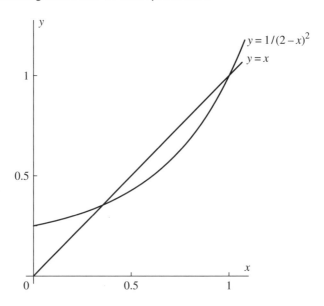

Fig. 9.2 *Graph showing the intersection of $y = x$ and $y = 1/(2 - x)^2$ in Example 9.3.*

From the previous work, the probability of ultimate extinction is the smallest positive root of

$$g = G(g) = \frac{1}{(2 - g)^2}.$$

Graphs of $y = x$ and $y = 1/(2 - x)^2$ are shown in Figure 9.2, which has one point of intersection for $0 < x < 1$, and the expected intersection at $(1, 1)$. The solution for g satisfies

$$g(2 - g)^2 = 1, \quad \text{or} \quad (g - 1)(g^2 - 3g + 1) = 0.$$

The required solution is

$$g = \frac{1}{2}(3 - \sqrt{5}) \approx 0.382,$$

which is the probability of ultimate extinction.

For $G(s)$,

$$\mu = G'(1) = 2, \qquad \sigma^2 = G''(1) + \mu - \mu^2 = 6 + 2 - 4 = 4.$$

Hence, by (9.7) and (9.12), the mean and variance of the nth generation are given by

$$\mu_n = \mu^n = 2^n, \qquad \sigma_n^2 = \sigma^2 n = 4n.$$

The mean, which behaves as 2^n, tends to infinity much faster than the variance, which behaves as n. The spread is not increasing as fast as the mean. □

Some general observations about the probability of ultimate extinction can be deduced from the probability generating function $G(s)$. Since

$$G(s) = \sum_{j=0}^{\infty} p_j s^j,$$

and $p_j \geq 0$ for all j, it follows that $G(s) \geq 0$ for $x \geq 0$. Term-by-term differentiation of the series also implies that

$$G'(s) \geq 0 \quad \text{and} \quad G''(s) \geq 0$$

for $s \geq 0$. The inequality $G'(s) \geq 0$ means that the slope of the curve $y = G(s)$ is never negative, and the inequality $G''(s) \geq 0$ implies that the slope is a monotonic increasing function of s, both for $s \geq 0$. A curve $y = G(s)$ (or simply the function $G(s)$) with the property $G''(s) \geq 0$ over any interval of s is said to be *convex* in s over the interval. Graphically, this means that the curve is always above its tangent for all values of s in the interval. The curve $y = G(s) = 1/(2-s)^2$ arising in Example 9.3 and shown in Figure 9.2 shows the convexity of the probability generating function.

Since $G(1) = 1$, the curve $y = G(s)$ always passes through the point $(1, 1)$. What is the condition that there is a further point of intersection between the curve and the line $y = s$ for $0 < s < 1$? The slope of the curve at $s = 1$ is $G'(1)$. Since the slope is monotonic increasing there can be no such point of intersection if $G'(1) \leq 1$. If $G'(1) > 1$ and $G(0) > 0$ then there will be an intermediate point of intersection. (If $G(0) = 0$, then the possibility of extinction does not arise.) Now $G'(1)$ is the mean μ of the probability distribution $\{p_j\}$. We conclude that the probability of extinction is certain if the mean $\mu \leq 1$, whilst if $\mu > 1$ then the population may become extinct at some future generation with probability whose value is the smallest positive root of $s = G(s)$.

9.5 Branching processes and martingales

Consider again the branching process starting with one individual so that $X_0 = 1$, and with $G(s)$ as the probability generating function of X_1. As before, X_n is the random variable representing the population size of the nth generation. Suppose we look at the expected value of the random variable X_{n+1} given the sequence of random variables $X_0, X_1, X_2, \ldots, X_n$. Now, by Section 1.10, such a conditional expectation is not a number but a *random variable*. Each set of descendants created by X_n will have mean $\mu = G'(1)$. Since there are X_n such means,

$$\mathbf{E}(X_{n+1}|X_0, X_1, X_2, \ldots, X_n) = X_n \mu. \tag{9.14}$$

Let

$$Z_n = \frac{X_n}{\mathbf{E}(X_n)} = \frac{X_n}{\mu^n}$$

using equation (9.7). The random variable Z_n is the random variable X_n *normalized* by its own mean since

$$E(Z_n) = E\left(\frac{X_n}{\mu^n}\right) = \frac{E(X_n)}{\mu^n} = 1 = E(X_0),$$

since the process starts with one individual in this case. Hence (9.14) becomes

$$E(Z_{n+1}\mu^{n+1}|X_1, X_2, \ldots, X_n) = Z_n\mu^{n+1},$$

or

$$E(Z_{n+1}|X_1, X_2, \ldots, X_n) = Z_n.$$

since $E(aX) = aE(X)$ where a is a constant and X is any random variable. Such a random variable sequence $\{Z_n\}$, in which the expected value of Z_{n+1} conditional on another random variable sequence $\{X_n\}$ is Z_n, is known as a *martingale*.

The most famous martingale arises in the following gambling problem. A gambler makes an even money bet with a casino starting with a pound bet. If she or he wins, the gambler has her or his £1 stake returned plus £1 from the casino. If the gambler loses the pound to the casino, she or he then bets £2, £4, £8,..., until she or he wins. Suppose the gambler first wins at the nth bet. Then the gambler will have won £2^n for an outlay of

$$£(1 + 2 + 2^2 + 2^3 + \cdots + 2^{n-1}) = £(2^n - 1),$$

summing the geometric series. Hence, the gambler always wins £1 at some stage. It is a guaranteed method of winning but does require a large financial backing, and for obvious reasons casinos do not permit this form of gambling. The martingale betting scheme will always beat what is a fair game in terms of the odds at each play.

We shall now explain why this gamble is a martingale. Suppose that the gambler starts with £1 and bets against the casino according to the rules outlined above. The doubling bets continue irrespective of the outcome at each bet. Let Z_n be the gambler's total asset or debt at the nth bet: $Z_0 = 1$ and Z_n can be a negative number indicating that the gambler owes money to the casino. Since $Z_0 = 1$, then Z_1 is a random variable that can take the values 0 or 2. Given Z_0 and Z_1, Z_2 can take any one of the values $-2, 0, 2, 4$, Z_3 can take any one of the values $-6, -4, -2, 0, 2, 4, 6, 8$, and so on as shown by the tree in Figure 9.3. The random variable Z_n can take any one of the values

$$\{-2^n + 2, -2^n + 4, \ldots, 2^n - 2, 2^n\}, \qquad (n = 1, 2, \ldots),$$

or, equivalently,

$$Z_n = \{-2^n + 2m + 2\} \qquad (m = 0, 1, 2, \ldots, 2^n - 1)$$

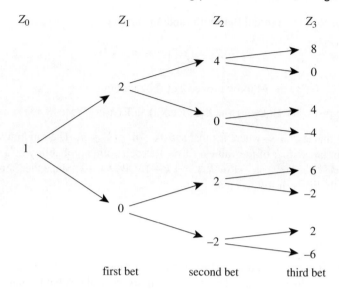

Z_0 Z_1 Z_2 Z_3

first bet   second bet   third bet

Fig. 9.3 *Possible outcomes of the first three bets in the gambling martingale.*

for $n = 1, 2, 3, \ldots$. The probability of any one of these values occurring at the nth bet is $1/2^n$. It follows that the expected value of Z_n (the gambler's mean asset) is

$$\mathbf{E}(Z_n) = \sum_{m=0}^{2^n-1} \frac{1}{2^n}(-2^n + 2m + 2) = 1,$$

after summing the arithmetic series. The expected winnings in a game of doubling bets would be £1, irrespective of the duration of the wager. Tactically, it would be best to play until the first win, and then start a new game with £1.

The conditional expectation of Z_3 given Z_0, Z_1 and Z_2 (remember these conditional expectations are random variables) is

$$\mathbf{E}(Z_3|Z_0, Z_1, Z_2) = \{-2, 0, 2, 4\} = Z_2,$$

as shown in Figure 9.3. The mean values of the pairs in the bottom row give the corresponding numbers in the previous row. Generally

$$\mathbf{E}(Z_{n+1}|Z_0, Z_1, Z_2, \ldots, Z_n) = Z_n,$$

which is a martingale.

Example 9.4. Let the random variable X_n be the position of a walker in a symmetric random walk after n steps. Suppose that the walk starts at the origin so that $X_0 = 0$. Show that $\{X_n\}$ is a martingale with respect to itself. (See Section 3.2 for a discussion of the properties of the symmetric random walk.)

In terms of the modified Bernoulli random variable

$$X_{n+1} = X_0 + \sum_{i=1}^{n+1} W_i = X_n + W_{n+1},$$

since $X_0 = 0$. By the Markov property of the random walk

$$\mathbf{E}(X_{n+1}|X_0, X_1, \ldots, X_n) = \mathbf{E}(X_{n+1}|X_n).$$

Suppose that $X_n = k$, where k must satisfy $-n \le k \le n$. The random variable W_{n+1} can take either of the values -1 or $+1$ each with a probability of $\frac{1}{2}$, so that $X_{n+1} = k-1$ if $W_{n+1} = -1$ or $X_{n+1} = k+1$ if $W_n = +1$. Since these outcomes are equally likely, it follows that

$$\mathbf{E}(X_{n+1}|X_n) = \frac{1}{2}(k-1) + \frac{1}{2}(k+1) = k = X_n,$$

which defines a martingale with respect to itself. $\qquad\square$

Example 9.5. Let $X_1, X_2, \ldots$ be independent random variables each of which can take the values -1 and $+1$ with probabilities of $\frac{1}{2}$. Show that the partial sums

$$Z_n = \sum_{j=1}^{n} \frac{1}{j} X_j, \quad (n = 1, 2, \ldots),$$

form a martingale with respect to $\{X_n\}$.

It follows that

$$\begin{aligned}
\mathbf{E}(Z_{n+1}|X_1, X_2, \ldots, X_n) &= \mathbf{E}(Z_n + \frac{1}{n+1} X_{n+1}|X_0, X_1, \ldots, X_n) \\
&= \mathbf{E}(Z_n|X_0, X_1, \ldots, X_n) \\
&\quad + (n+1)\mathbf{E}(X_{n+1}|X_0, X_1, \ldots, X_n) \\
&= Z_n + (n+1)\mathbf{E}(X_{n+1}) = Z_n,
\end{aligned}$$

since $\mathbf{E}(X_{n+1}) = 0$.

The random variable Z_n is the partial sum of a harmonic series with terms $1/j$ with randomly selected signs, known as a random harmonic series. It is known that the harmonic series

$$\sum_{j=1}^{\infty} \frac{1}{j} = 1 + \frac{1}{2} + \frac{1}{3} + \frac{1}{4} + \cdots$$

diverges and that the alternating harmonic series

$$\sum_{j=1}^{\infty} (-1)^{j+1} \frac{1}{j} = 1 - \frac{1}{2} + \frac{1}{3} - \frac{1}{4} + \cdots$$

converges. It can be shown that the random harmonic series converges with probability 1. This requires considerable extra theory: a full account can be found in Lawler (1995). $\qquad\square$

9.6 Stopping rules

Suppose we have a random process, say a Markov chain or a branching process, with random variables $\{X_r\}$, $(r = 0, 1, 2, \ldots)$ in which we decide to stop the process when a specified condition is met. Hence $\{X_r\}$ may go though a sequence of values $(x_0, x_1, \ldots, x_n)$ and then stop at x_n when a predetermined stopping condition is met at x_n. The random variable T for the *stopping time* or number of steps can take any of the integer values $(0, 1, 2, \ldots)$. A simple example of a stopping rule is that $n = m$ where m is a given positive integer: in other words, the process stops after m steps. In this case, just given T, the solution is known but X_m is not, in which case the expected value $\mathbf{E}(X_m)$ will be of interest. Alternatively, we might wish to find the expected value $\mathbf{E}(T)$ for a problem in which the stopping rule is that the process stops when $X_n \geq x$, where x is a given number.

Example 9.6. A random walk starts at the origin, with probability p that the walk advances one step and probability $q = 1 - p$ that the walk retreats one step at every position. The walk stops after $T = 10$ steps. What is the expected position of the walk at the stopping time?

In terms of the modified Bernoulli random variable $\{W_i\}$ (see Section 3.2), the random variable of the position of the walk after n steps is

$$X_n = X_0 + \sum_{i=1}^{n} W_i = \sum_{i=1}^{n} W_i$$

since $X_0 = 0$. It was shown in Section 3.2, that

$$\mathbf{E}(X_n) = n(p - q) = n(2p - 1).$$

Hence, for $T = 10$, the expected position is

$$\mathbf{E}(X_{10}) = 10(p - q).$$

$\square$

Another stopping rule could be that the process stops when the random variable X_n first takes a particular value or several values. This could be a particular strategy in a gambling game where the gambler abandons the game when certain winnings are achieved or cuts his/her losses when a certain deficit is reached.

The following random walk is a version of the gambler's ruin problem of Section 2.2.

Example 9.7. A symmetric random walk starts at position k and ends when the walk first reaches either the origin or position a, where k and a are positive integers such that $(0 < k < a)$. Let T be the stopping time, that is, the number of steps to achieve this. If X_n is the position of the walk after n steps, show that $\mathbf{E}(X_T) = k$.

Formally we can write the stopping time T as

$$T = \min\{n : X_n = 0 \text{ or } X_n = a\},$$

which means that T is the minimum value of n at which the walk first visits either the origin or a: the boundaries 0 and a are absorbing states. This problem is also recognizable as a symmetric gambler's ruin: the gambler is ruined when the walk reaches the origin. In Section 2.2, we showed that the probability that the origin is reached first is $(a - k)/a$ (see equation (2.5)). The probability that a is reached first is its complement k/a. Hence, the expected value of X_T is given by

$$\mathbf{E}(X_T) = 0. \left(\frac{a - k}{a}\right) + a. \left(\frac{k}{a}\right) = k,$$

as required. □

We showed in Example 9.4 that $\{X_n\}$ is a martingale with respect to the modified Bernoulli random variable $\{W_n\}$. It has just been shown that

$$\mathbf{E}(X_T) = k = X_0 = \mathbf{E}(X_0),$$

since X_0 is specified in this problem. For this martingale, and for certain other martingales generally, this result is not accidental. In fact the *stopping theorem* states:

If $\{Z_n\}$ is a martingale with respect to $\{X_n\}$ $(n = 0, 1, 2, \ldots)$, T is a defined stopping time, and

(i) $\mathbf{P}(T < \infty) = 1$,
(ii) $\mathbf{E}(|Z_T|) < \infty$,
(iii) $\mathbf{E}(Z_n | T > n)\mathbf{P}(T > n) \to 0$ as $n \to \infty$,

then $\mathbf{E}(Z_T) = \mathbf{E}(Z_0)$.

This useful result, which we shall not prove here (see Grimmett and Stirzaker, 1982, p. 209, or Lawler, 1995, p. 93), interpreted for the example above, asserts that the expected position of the walk that finishes at either 0 or a is simply its initial position k.

The following example shows the link between another martingale associated with the symmetric random walk and the gambler's ruin problem and the expected duration of the game and the stopping time T. We shall simply verify the result without establishing the conditions in the stopping theorem.

Example 9.8. Consider again the symmetric random walk described in Example 9.7, with stopping time T defined in the same manner. Show that the random variable $\{Y_n\}$, $(n = 0, 1, 2, \ldots)$ where $Y_n = X_n^2 - n$ is also a martingale with respect to $\{X_n\}$. Show that

$$\mathbf{E}(Y_T) = \mathbf{E}(Y_0) = k^2.$$

To establish that $\{Y_n\}$ is a martingale we must show that

$$\mathbf{E}(Y_{n+1}|X_0, X_1, X_2, \ldots, X_n) = Y_n.$$

Since the walk is a Markov process,

$$\mathbf{E}(Y_{n+1}|X_0, X_1, X_2, \ldots, X_n) = \mathbf{E}(Y_{n+1}|X_n) = \mathbf{E}(X_{n+1}^2 - (n+1)|X_n)$$
$$= X_n^2 + 1 - (n+1) = Y_n,$$

The latter random variable follows because

$$\mathbf{E}(X_{n+1}^2|X_n) = \frac{1}{2}(X_n + 1)^2 + \frac{1}{2}(X_n - 1)^2 = X_n^2 + 1,$$

and

$$\mathbf{E}(n+1|X_n) = n + 1.$$

Hence $\{Y_n\}$ is a martingale with respect to $\{X_n\}$.

If $n = T$, the stopping time, then

$$\mathbf{E}(Y_T) = \mathbf{E}(X_T^2) - \mathbf{E}(T)$$
$$= 0.\mathbf{P}(X_T = 0) + a^2.\mathbf{P}(X_T = a) - \mathbf{E}(T)$$
$$= a^2.\frac{k}{a} - \mathbf{E}(T) = ak - \mathbf{E}(T).$$

Now $\mathbf{E}(T)$ is the expected duration of the symmetric gambler's ruin problem, and it was shown in Section 2.4, equation (2.10), that

$$\mathbf{E}(T) = d_k = k(a - k).$$

Hence

$$\mathbf{E}(Y_T) = ak - k(a - k) = k^2 = \mathbf{E}(Y_0).$$

Usually, the procedure is used in reverse assuming that the theorem has been proved and is available. In that case the theorem then implies that $\mathbf{E}(Y_T) = \mathbf{E}(Y_0) = k^2$, and the expected stopping time or duration $\mathbf{E}(T)$ can be deduced.

□

9.7 The simple epidemic

The subject of epidemiology from a theoretical viewpoint is concerned with the construction of stochastic models that can represent the spread of a specific disease, through a population. It is a topic of particular current interest with the apparent advent of new virulent diseases, which in some cases have no curative treatment. Some diseases are cyclical, such as measles and influenza; some diseases are currently spreading, such as HIV infections, which have long periods in which infected persons can transmit the disease to others; some diseases have disappeared, such as smallpox; and some, which have declined through treatment, such as tuberculosis, are recurring through drug-resistant strains. The main aim of probability models is to be able to predict the likely extent of a disease – how many are infected though the period of the epidemic and how might inoculation affect the spread of the disease? There is also interest in the geographical, demographic and behaviour over time of diseases.

In this section we shall develop a model for the *simple epidemic*. In epidemiology there are assumed to be three groups of individuals in a population. These are the *susceptibles*, those who are might succumb to the disease, the *infectives*, those that have the disease and can spread it among the susceptibles, and what we might lump together as the *immunes*, which includes, in addition to those who are immune, the isolated, the dead (as a result of the disease) and the recovered and now immune.

In practice, the spread of diseases is more complcated than simple stochastic models display. They can be affected by many factors. Here is a fairly general description of the progress of a typical disease. If someone becomes infected there is usually a *latent period* during which the disease develops. At the end of this period the patient might display symptoms and become an *infective*, although the two events might not occur at the same time. The individual remains an infective during the *infectious period*, after which the patient either dies or recovers and perhaps becomes immune to further bouts of the illness.

In the simple epidemic the assumptions are very restrictive. It is assumed that the population contains only susceptibles and infectives – individuals do not die or recover but remain infective. This model is not very realistic but it might cover the early stages of a disease that has a long infectious period. Suppose that the population remains fixed at $n_0 + 1$ with no births or deaths during the epidemic and that one infective is introduced at time $t = 0$. Initially there are n_0 susceptibles. Let the random variable $X(t)$ represent the number of susceptibles that are *not* infected at time t. At this time there will be $n_0 + 1 - X(t)$ infectives. It is assumed that individuals in the population mix homogeneously, which is rarely the case – except possibly with animals under controlled laboratory conditions. In the simple epidemic, the main assumption is that the likelihood of an infection occurring in a time interval δt is

$$\beta X(t)[n_0 + 1 - X(t)]\delta t,$$

that is, it is proportional to the product of the numbers of susceptibles and infectives at time t. In this joint dependence, the likelihood of infection will be high if both populations are relatively large, and small if they are both small. The constant β is known as the *contact rate*. The probability of more than one infection taking place in time δt is assumed to be negligible. Let $p_n(t)$, $0 \le n \le n_0$ be the probability that there are n susceptibles at time t. Then, by the partition law,

$$p_n(t + \delta t) = \beta(n + 1)(n_0 - n)\delta t p_{n+1}(t) + [1 - \beta n(n_0 + 1 - n)\delta t]p_n(t) + o(\delta t),$$

for $0 \leq n \leq n_0 - 1$, and

$$p_{n_0}(t) = (1 - \beta n_0 \delta t) p_{n_0}(t) + o(\delta t).$$

In the first equation, either one susceptible became an infective in time δt with probability $\beta(n + 1)(n_0 - n)\delta t$ when the population was $n + 1$ at time t, or no infection occurred with probability $1 - \beta n(n_0 + 1 - n)\delta t$ when the population was n. Following the usual method of dividing though by δt and then letting $\delta t \to 0$, we arrive at the differential-difference equations for the simple epidemic:

$$\frac{dp_n(t)}{dt} = \beta(n + 1)(n_0 - n)p_{n+1}(t) - \beta n(n_0 + 1 - n)p_n(t), \qquad (9.15)$$

for $n = 0, 1, 2, \ldots n_0 - 1$ and

$$\frac{dp_{n_0}(t)}{dt} = -\beta n_0 p_{n_0}(t). \qquad (9.16)$$

Since there are n_0 susceptibles at time $t = 0$, the initial condition must be $p_{n_0}(0) = 1$ with $p_n(0) = 0$ for $n = 0, 1, 2, \ldots, n_0 - 1$.

9.8 An iterative solution scheme for the simple epidemic

A partial differential equation for the probability generating function for equations (9.15) and (9.16) can be obtained (see Bailey, 1964), but the equation has no simple solution for the epidemic problem. Here we show how solutions for $p_n(t)$ can be constructed by an iterative procedure for any number of probabilities, although the solutions do become increasing complicated and they do not suggest an obvious general formula. It is, however, possible to obtain explicit formulae for epidemics in very small populations.

We first eliminate the parameter β by letting $\tau = \beta t$, by putting

$$p_n(\tau/\beta) = q_n(\tau) \qquad (9.17)$$

so that eqns (9.15) and (9.16) become

$$\frac{dq_n(\tau)}{d\tau} = (n + 1)(n_0 - n)q_{n+1}(\tau) - n(n_0 + 1 - n)q_n(\tau), \qquad (0 \leq n \leq n_0 - 1),$$
$$\qquad (9.18)$$

$$\frac{dq_{n_0}(\tau)}{d\tau} = -n_0 q_{n_0}(\tau). \qquad (9.19)$$

Now write the equations as (it is now convenient to take (9.19) first)

$$\frac{d}{d\tau}\left[e^{n_0 \tau} q_{n_0}(\tau)\right] = 0,$$

$$\frac{d}{d\tau}\left[e^{n(n_0+1-n)\tau}q_n(\tau)\right] = (n+1)(n_0-n)e^{n(n_0+1-n)\tau}q_{n+1}(\tau).$$

Now let

$$u_n(\tau) = e^{n(n_0+1-n)\tau}q_n(\tau). \tag{9.20}$$

Then $u_n(\tau)$, $n = 0, 1, 2, \ldots, n_0$ satisfy the differential equations

$$\frac{du_{n_0}(\tau)}{d\tau} = 0, \tag{9.21}$$

$$\frac{du_n(\tau)}{d\tau} = (n+1)(n_0-n)e^{(2n-n_0)\tau}u_{n+1}(\tau), \tag{9.22}$$

where the initial conditions are $u_{n_0}(0) = 1$ and $u_n = 0$ for $n \neq n_0$. It follows from equation (9.21) that

$$u_{n_0}(\tau) = 1. \tag{9.23}$$

From equation (9.22), the equation for $u_{n_0-1}(\tau)$ is

$$\frac{du_{n_0-1}(\tau)}{d\tau} = n_0 e^{(n_0-2)\tau}u_{n_0}(\tau) = n_0 e^{(n_0-2)\tau}.$$

Hence

$$u_{n_0-1}(\tau) = \frac{n_0}{n_0-2}\left[e^{(n_0-2)\tau} - 1\right]. \tag{9.24}$$

Using this solution for $u_{n_0-1}(\tau)$, the equation for $u_{n_0-2}(\tau)$ is

$$\begin{aligned}
\frac{du_{n_0-2}(\tau)}{d\tau} &= 2(n_0-1)e^{(n_0-4)\tau}u_{n_0-1}(\tau) \\
&= 2(n_0-1)e^{(n_0-4)\tau}\frac{n_0}{n_0-2}\left[e^{(n_0-2)\tau} - 1\right] \\
&= \frac{2.n_0(n_0-1)}{n_0-2}\left[e^{2(n_0-3)\tau} - e^{(n_0-4)\tau}\right].
\end{aligned} \tag{9.25}$$

Routine integration, together with the initial condition, gives (for $n_0 > 4$)

$$u_{n_0-2}(\tau) = \frac{n_0(n_0-1)\left[(n_0-4)\{e^{2(n_0-3)\tau} - 1\} - 2(n_0-3)\{e^{(n_0-4)\tau} - 1\}\right]}{(n_0-2)(n_0-3)(n_0-4)}.$$

This procedure can be continued but the number of terms increases at each step. The probabilities can be recovered from equations (9.17) and (9.19).

Example 9.9. One infective is living with a group of four individuals who are susceptible to the disease. Treating this situation as a simple epidemic in an isolated group, find the probability that all members of the group have contracted the disease at time t.

In the notation above we require $p_4(t)$. In the iteration scheme $n_0 = 4$ so that from equations (9.23) and (9.24)

$$u_4(\tau) = 1, \qquad u_3(\tau) = 2(e^{2\tau} - 1).$$

However, there is a change in equation (9.25) if $n_0 = 4$. The differential equation becomes

$$\frac{d}{d\tau}u_2(\tau) = 12(e^{2\tau} - 1),$$

which, subject to $u_2(0) = 0$, has the solution

$$u_2(\tau) = -6(1 + 2\tau) + 6e^{2\tau}.$$

The equation for $u_1(\tau)$ is

$$\frac{d}{d\tau}u_1(\tau) = 36 - 36(1 + 2\tau)e^{-2\tau},$$

which has the solution

$$u_1(\tau) = 36(\tau - 1) + 36(1 + \tau)e^{-2\tau}.$$

A similar process gives the last function:

$$u_0(\tau) = 1 + (27 - 36\tau)e^{-4\tau} - (28 + 24\tau)e^{-6\tau}.$$

From equations (9.17) and (9.20) the probabilities can be now be found:

$$p_4(t) = e^{-4\beta t}, \quad p_3(t) = 2e^{-4\beta t} - e^{-6\beta t}, \quad p_2(t) = 6e^{-4\beta t} - (6 + 12\beta t)e^{-6\beta t},$$

$$p_1(t) = 36(\beta t - 1)e^{-4\beta t} + 36(1 + \beta t)e^{-6\beta t},$$

$$p_0(t) = 1 + 9(3 - 4\beta t)e^{-4\beta t} - 4(7 + 6\beta t)e^{-6\beta t}.$$

A useful check on these probabilities is that they must satisfy

$$\sum_{n=0}^{4} p_n(t) = 1.$$

The formula for $p_0(t)$ gives the probability that all individuals have contracted the disease by time t. Note that $p_0(t) \to 1$ as $t \to \infty$, which means that all susceptibles ultimately catch the disease, a characteristic of the simple epidemic.

□

Problems

9.1. In a branching process, the probability that any individual has j descendants is given by

$$p_0 = 0, \qquad p_j = \frac{1}{2^j}, \quad (j \geq 1).$$

Show that the probability generating function of the first generation is

$$G(s) = \frac{s}{2-s}.$$

Find the further generating functions $G_2(s)$, $G_3(s)$ and $G_4(s)$. Show by induction that

$$G_n(s) = \frac{s}{2^n - (2^n - 1)s}.$$

Find $p_{n,j}$, the probability that the population size of the nth generation is j given that the process starts with one individual. What is the mean population size of the nth generation?

9.2. Suppose in a branching process that any individual has a probability

$$p_j = (1 - p)p^j, \qquad (j = 0, 1, 2, \ldots),$$

of producing j descendants in the next generation, where p $(0 < p < 1)$ is a constant. Find the probability generating function of the second and third generations. What is the mean size of any generation?

9.3. A branching process has the probability generating function

$$G(s) = a + bs + (1 - a - b)s^2$$

for the descendants of any individual, where a and b satisfy the inequalities

$$1 > a > 0, \quad b > 0, \quad a + b < 1.$$

Given that the process starts with one individual, discuss the nature of the descendant generations. What is the maximum possible size of the nth generation? Show that extinction in the population is certain if $2a + b \geq 1$.

9.4. A branching process starts with one individual. Subsequently, any individual has a probability

$$p_j = \frac{\lambda^j e^{-\lambda}}{j!}, \qquad (j = 0, 1, 2, \ldots)$$

of producing j descendants. Find the probability generating function of this distribution. Obtain the mean and variance of the nth generation. Show that the probability of ultimate extinction is certain if $\lambda \leq 1$.

9.5. A branching process starts with one individual. Any individual has a probability

$$p_j = \frac{\lambda^{2j} \operatorname{sech} \lambda}{(2j)!}, \qquad (j = 0, 1, 2, \ldots)$$

of producing j descendants. Find the probability generating function of this distribution. Obtain the mean size of the nth generation. Show that ultimate extinction is certain if $\lambda \tanh \lambda \leq 1$.

9.6. A branching process starts with two individuals. Either individual, and any of their descendants, has probability p_j, $(j = 0, 1, 2, \ldots)$ of producing j descendants independently of any other individual. Explain why the probabilities of $0, 1, 2, \ldots$ descendants in the first generation are

$$p_0^2, \quad p_0 p_1 + p_1 p_0, \quad p_0 p_2 + p_1 p_1 + p_2 p_0, \quad \cdots \quad \sum_{i=0}^{n} p_i p_{n-i}, \ldots,$$

respectively. Hence show that the probability generating function of the first generation is $G(s)^2$, where

$$G(s) = \sum_{j=0}^{\infty} p_j s^j.$$

The second generation from each original individual has generating function $G_2(s) = G(G(s))$ (see Section 9.2). Explain why the probability generating function of the second generation is $G_2(s)^2$, and that of the nth generation is $G_n(s)^2$.

If the branching process starts with r individuals, what would you think is the formula for the probability generating function of the nth generation?

9.7. A branching process starts with two individuals, as in the previous problem. The probability

$$p_j = \frac{1}{2^{j+1}}, \qquad (j = 0, 1, 2, \ldots).$$

Using the results from Example 9.1, find $H_n(s)$, the probability generating function of the nth generation. Find also

(a) the probability that the size of the population of the nth generation is m;

(b) the probability of extinction by the nth generation;

(c) the probability of ultimate extinction.

9.8. A branching process starts with r individuals, and each individual produces descendants with probability distribution $\{p_j\}$, $(j = 0, 1, 2, \ldots)$, which has the probability generating function $G(s)$. Given that the probability of the nth generation is $[G_n(s)]^r$, where $G_n(s) = G(G(\ldots(G(s))\ldots))$, find the mean population size of the nth generation in terms of $\mu = G'(1)$.

9.9. Let X_n be the random variable of the population size of a branching process starting with one individual. Suppose that all individuals survive, and that

$$Z_n = 1 + X_1 + X_2 + \cdots + X_n$$

is the random variable of the accumulated population with $\mu = \mathbf{E}(X_1)$.

(a) If H_n is the probability generating function of the total accumulated population, Z_n, up to and including the nth generation, show that (see Problem 1.18)

$$H_1(s) = sG(s), \qquad H_2(s) = sG(H_1(s)) = sG(sG(s)),$$

(which perhaps gives a clue to the probability generating function of $H_n(s)$),

(b) What is the mean accumulated population size $\mathbf{E}(Z_n)$ (you do not require $H_n(s)$ for this formula)?

(c) If $\mu < 1$, what is $\lim_{n \to \infty} \mathbf{E}(Z_n)$, the ultimate expected population?

(d) What is the variance of Z_n?

9.10. A branching process starts with one individual and each individual has probability p_j of producing j descendants independently of every other individual. Find the mean and variance of $\{p_j\}$ in each of the following cases, and hence find the mean and variance of the population of the nth generation:

(a) $$p_j = \frac{e^{-\mu}\mu^j}{j!}, \qquad (j = 0, 1, 2, \ldots) \quad \text{(Poisson)};$$

(b) $$p_j = (1 - p)^{j-1}p \qquad (j = 1, 2, \ldots; 0 < p < 1) \quad \text{(geometric)};$$

(c) $$p_j = \binom{r + j - 1}{r - 1} p^j (1 - p)^r,$$

$$(j = 0, 1, 2, \ldots; 0 < p < 1) \quad \text{(negative binomial)},$$

where r is a positive integer.

9.11. A branching process has a probability generating function

$$G(s) = \left(\frac{1 - p}{1 - ps}\right)^r,$$

where r is a positive integer, the process having started with one individual (a negative binomial distribution). Show that extinction is not certain if $p > 1/(1+r)$. Find the probability of extinction if $r = 2$ and $p > \frac{1}{3}$.

9.12. Let $G_n(s)$ be the probability generating function of the population size of the nth generation of a branching process. The probability that the population size is zero at the nth generation is $G_n(0)$. What is the probability that the population actually becomes extinct *at* the nth generation?

In Example 9.1, where $p_j = 1/2^{j+1}$ $(j = 0, 1, 2, \ldots)$, it was shown that

$$G_n(s) = \frac{n}{n+1} + \sum_{r=1}^{\infty} \frac{n^{r-1}}{(n+1)^{r+1}} s^r.$$

Find the probability of extinction,

(a) at the nth generation,

(b) at the nth generation or later.

What is the mean number of generations until extinction occurs?

9.13. An annual plant produces n seeds in a season which are assumed to have a Poisson distribution with parameter λ. Each seed has a probability p of germinating to create a new plant that propagates in the following year. Let N be the random variable of the number of seeds, and M the random variable of the number of new plants. Show that p_m, the probability that there are m growing plants in the first year is given by

$$p_m = (p\lambda)^m e^{-p\lambda}/m!,$$

that is Poisson with parameter $p\lambda$ (see also Problem 1.20). Show that its probability generating function is

$$G(s) = e^{p\lambda(s-1)}.$$

Assuming that all the germinated plants survive and that each propagates in the same manner in succeeding years, find the mean number of plants in year k. Show that extinction is certain if $p\lambda \leq 1$.

9.14. The version of Example 9.1 with a geometric distribution is the branching process with $p_j = (1-p)p^j$, $(0 < p < 1; \; j = 0, 1, 2, \ldots)$. Show that

$$G(s) = \frac{1-p}{1-ps}.$$

Using an induction method, prove that

$$G_n(s) = \frac{(1-p)[p^n - (1-p)^n - ps\{p^{n-1} - (1-p)^{n-1}\}]}{[p^{n+1} - (1-p)^{n+1} - ps\{p^n - (1-p)^n\}]}, \quad (p \neq \tfrac{1}{2}).$$

Find the mean and variance of the population size of the nth generation.

What is the probability of extinction by the nth generation? Show that ultimate extinction is certain if $p < \tfrac{1}{2}$, but has probability $(1-p)/p$ if $p > \tfrac{1}{2}$.

9.15. A branching process starts with one individual, and the probability of producing j descendants has the distribution $\{p_j\}$, $(j = 0, 1, 2, \ldots)$. The same probability distribution applies independently to all descendants and their descendants. If X_n is the random variable of the size of the nth generation, show that

$$\mathbf{E}(X_n) \geq 1 - \mathbf{P}(X_n = 0).$$

In Section 9.3 it was shown that $\mathbf{E}(X_n) = \mu^n$, where $\mu = \mathbf{E}(X_1)$. Deduce that the probability of extinction eventually is certain if $\mu < 1$.

9.16. In a branching process starting with one individual, the probability that any individual has j descendants is $p_j = \alpha/2^j$, $(j = 0, 1, 2, \ldots, R)$, where α is a constant. This means that any individual can have a maximum of R descendants. Find α and the probability generating function $G(s)$ of the first generation. Show that the mean size of the nth generation is

$$\mu_n = \left[\frac{2^{R+1} - 2 - R}{2^{R+1} - 1} \right]^n.$$

What is the probability of ultimate extinction?

9.17. Extend the tree in Figure 9.3 for the gambling martingale in Section 9.5 to Z_4, and confirm that

$$\mathbf{E}(Z_4 | Z_0, Z_1, Z_2, Z_3) = Z_3.$$

Confirm also that $\mathbf{E}(Z_4) = 1$.

9.18. A gambling game similar to the gambling martingale of Section 9.5 is played according to the following rules:

(a) the gambler starts with £1, but has unlimited resources;
(b) against the casino, which also has unlimited resources, the gambler plays a series of games in which the probability that the gambler wins is $1/p$ and loses is $(p-1)/p$, where $p > 1$;
(c) at the nth game, the gambler either wins £$(p^n - p^{n-1})$ or loses £p^{n-1}.

If Z_n is the random variable of the gambler's asset/debt at the nth game, draw a tree diagram similar to that of Figure 9.3 as far as Z_3. Show that

$$Z_3 = \{-p - p^2, -p^2, -p, 0, p^3 - p^2 - p, p^3 - p^2, p^3 - p, p^3\}$$

and confirm that

$$\mathbf{E}(Z_2 | Z_0, Z_1) = Z_1, \qquad \mathbf{E}(Z_3 | Z_0, Z_1, Z_2) = Z_2,$$

which indicates that this game is a martingale. Show also that

$$\mathbf{E}(Z_1) = \mathbf{E}(Z_2) = \mathbf{E}(Z_3) = 1.$$

Assuming that it is a martingale, show that, if the gambler first wins at the nth game, then the gambler will have an asset gain or debt of $£(p^{n+1} - 2p^n + 1)/(p-1)$. Explain why a win for the gambler can only be guaranteed for all n if $p \geq 2$.

9.19. Let $X_1, X_2, \ldots$ be independent random variables with means $\mu_1, \mu_2, \ldots$ respectively. Let

$$Z_n = X_1 + X_2 + \cdots + X_n,$$

and let $Z_0 = X_0 = 0$. Show that the random variable

$$Y_n = Z_n - \sum_{i=1}^{n} \mu_i, \quad (n = 1, , 2, \ldots)$$

is a martingale with respect to $\{X_n\}$. [Note that $\mathbf{E}(Z_{n+1}|X_1, X_2, \ldots, X_n) = Z_n$.]

9.20. Consider an unsymmetric random walk which starts at the origin. The walk advances one position with probability p and retreats one position with probability $1 - p$. Let X_n be the random variable giving the position of the walk at step n. Let Z_n be the random variable given by

$$Z_n = X_n + (1 - 2p)n.$$

Show that

$$\mathbf{E}(Z_2|X_0, X_1) = \{-2p, 2 - 2p\} = Z_1.$$

Generally, show that $\{Z_n\}$ is a martingale with respect to $\{X_n\}$.

9.21. In the gambling martingale of Section 9.5, the random variable Z_n, the gambler's asset, in a game against a casino in which the gambler starts with £1 and doubles the bid at each play, is given by

$$Z_n = \{-2^n + 2m + 2\}, \quad (m = 0, 1, 2, \ldots, 2^n - 1).$$

Find the variance of Z_n. What is the variance of

$$\mathbf{E}(Z_n|Z_0, Z_1, \ldots, Z_{n-1})?$$

9.22. A random walk starts at the origin, and, with probability p_1, advances one position, and with probability $q_1 = 1 - p_1$ retreats one position, at every step. After 10 steps the probabilities change to p_2 and $q_2 = 1 - p_2$ respectively. What is the expected position of the walk after a total of 20 steps?

9.23. A symmetric random walk starts at the origin $x = 0$. The stopping rule that the walk ends when the position $x = 1$ is first reached is applied; that is, the stopping time T is given by

$$T = \min\{n : X_n = 1\},$$

where X_n is the position of the walk at step n. What is the expected value of T? If this walk was interpreted as a gambling problem in which the gambler starts with nothing, with equal odds of winning or losing £1 at each play, what is the flaw in this stopping rule as a strategy of guaranteeing a win for the gambler in every game? [Hint: the generating function for the probability of the first passage is

$$G(s) = [1 - (1 - s^2)^{\frac{1}{2}}]/s :$$

see Problem 3.11.]

9.24. In a finite-state branching process, the descendant probabilities are, for every individual,

$$p_j = \frac{2^{m-j}}{2^{m+1} - 1}, \qquad (j = 0, 1, 2, \ldots, m),$$

and the process starts with one individual. Find the mean size of the first generation. If X_n is a random variable of the size of the nth generation, explain why

$$Z_n = \left[\frac{2^{m+1} - 1}{2^{m+1} - m - 2}\right]^n X_n$$

defines a martingale over $\{X_n\}$.

9.25. A random walk starts at the origin, and at each step the walk advances one position with probability p or retreats with probability $1 - p$. Show that the random variable

$$Y_n = X_n^2 + 2(1 - 2p)nX_n + [(2p - 1)^2 - 1]n + (2p - 1)^2 n^2,$$

where X_n is the random variable of the position of the walk at time n, defines a martingale with respect to $\{X_n\}$. Show also that

$$\mathbf{E}(Y_n) = \mathbf{E}(Y_1) = 0,$$

which is a characteristic result for martingales.

9.26. A simple epidemic has n_0 susceptibles and one infective at time $t = 0$. If $p_n(t)$ is the probability that there are n susceptibles at time t, it was shown in Section 9.7 that $p_n(t)$ satisfies the differential-difference equations (see equations (9.15) and (9.16))

$$\frac{\mathrm{d}p_n(t)}{\mathrm{d}t} = \beta(n + 1)(n_0 - n)p_{n+1}(t) - \beta n(n_0 + 1 - n)p_n(t),$$

for $n = 0, 1, 2, \ldots n_0 - 1$ and

$$\frac{\mathrm{d}p_{n_0}(t)}{\mathrm{d}t} = -\beta n_0 p_{n_0}(t).$$

Show that the probability generating function

$$G(s, t) = \sum_{n=0}^{n_0} p_n(t) s^n$$

satisfies the partial differential equation

$$\frac{\partial G}{\partial t} = \beta(1 - s) \left[n_0 \frac{\partial G}{\partial s} - s \frac{\partial^2 G}{\partial s^2} \right].$$

9.27. In a simple epidemic with initially n_0 susceptibles and one infective, the probability generating function $G(s, t)$ for the infective population is given by

$$\frac{\partial G}{\partial t} = \beta(1 - s) \left[n_0 \frac{\partial G}{\partial s} - s \frac{\partial^2 G}{\partial s^2} \right],$$

(see Problem 9.26). Non-dimensionalize the equation by putting $\tau = \beta t$. For small τ let

$$G(s, \tau/\beta) = G_0(s) + G_1(s)\tau + G_2(s)\tau^2 + \cdots.$$

Show that

$$n G_n(s) = n_0(1 - s) \frac{\partial G_{n-1}(s)}{\partial s} - s(1 - s) \frac{\partial^2 G_{n-1}(s)}{\partial s^2},$$

for $n = 1, 2, 3, \ldots n_0$. What is $G_0(s)$? Find the coefficients $G_1(s)$ and $G_2(s)$. Hence show that the mean number of infectives for small τ is given by

$$n_0 - n_0\tau - \frac{1}{2} n_0(n_0 - 2)\tau^2 + O(\tau^3).$$

In Example 9.9, the number of susceptibles is initially given by $n_0 = 4$. Expand $p_0(t)$, $p_1(t)$ and $p_2(t)$ in powers of τ and confirm that the expansions agree with $G_1(s)$ and $G_2(s)$ above.

10
Computer simulations and projects

The software *Mathematica* has been used extensively in this book to confirm answers in worked examples and in the problems at the end of chapters, and in generating graphs. Symbolic computation systems such as *Mathematica*, *Maple*, *MATLAB*, etc, are being introduced into many courses in Mathematics, Statistics, Engineering and the Physical Sciences. It is seen as a necessary training in the use of computers in these subjects. Readers of this book are encouraged to sample some of the projects listed below in conjunction with the text.

The following projects follow the chapters in the main text. They are specifically designed so that they do not require elaborate programming in *Mathematica*, and use mainly built-in *Mathematica* commands. It should be emphasized that these computer programs generally generate answers without explaining just how any results were obtained. Symbolic computing is not yet a substitute for understanding the theory behind stochastic processes. *Mathematica* notebooks for the projects are available at the Keele University, Mathematics Department website:

http://www.keele.ac.uk/depts/ma/

The programs are intended to be flexible in that inputs, parameters, data, etc. can be varied. They are intended to give ideas for other programs. Not all topics in the book are covered, but many programs can be adapted to other applications. Computations and graphs are often interesting in that they give some idea of scale and importance in problems: a result can look interesting as a formula but be of negligible consequence numerically. The projects should only be attempted in conjunction with the programs: pen and paper analysis is not recommended.

Chapter 1: Some background in probability

10-1.1. Two distinguishable dice are rolled. Design a program to simulate the sum of the two face values shown for 1000, say, random throws. Compare the average of these values with the theoretical expected sum of 7, as in Problem 1.11. Compare also the theoretical variance of the sum with the simulated value.

10-1.2. Write a computer program to simulate the case of three dice that are rolled and their face values noted. By running the program, say, 2000 times compare by calculating the ratio of the number of cases in which two dice show the same face value and the third is different with the theoretical probability of 5/12. Estimate the probability that the three dice all show different values with the theoretical probability.

10-1.3. A fair coin is spun n times, and the sequence of faces shown recorded. Write a program to simulate this Bernoulli experiment (see Section 1.7). Count the frequencies of the occurrences of heads and tails over, say, $n = 1000$ trials.

Continue the program by counting in a simulation the number of times, say, a head appears in say 20 spins, and perform the simulation $m = 2000$ times. Compare the data obtained with the exact binomial distribution that the theory predicts for this problem.

10-1.4. The gamma distribution has the density function

$$f(x) = \frac{\alpha^n}{\Gamma(n)} x^{n-1} e^{-\alpha x}.$$

Devise a program to show graphically by surfaces the cumulative and density functions for $n = 2$, say, in terms of x and the parameter $\lambda = 1/\alpha$ (*Mathematica* uses the parameter λ).

Chapter 2: Some gambling problems

10-2.1. Write a program to simulate the gambler's ruin problem in which a gambler with k units bets against an opponent with $(a - k)$ units with a probability p of the gambler winning at each play. Show a graph of the gambler's current stake against the number of the play, such that the graph terminates when either of the absorbing states at $k = 0$ or $k = a$ is reached. See Figure 2.1 in Section 2.3.

10-2.2. In a single trial the possible score outcomes 1 and 2 can occur with probabilities α and $1 - \alpha$ where $0 < \alpha < 1$ (see Problem 2.2). A succession of such trials take place and each time the scores are accumulated from zero. Design a program to simulate this process, and count the number of times the total scores 1, 2, 3, ..., m occur. Possible numbers are $\alpha = \frac{1}{4}$, $m = 20$, scores run 1000 times. The theory gives the probability that score n occurs as

$$p_n = \frac{1 - (\alpha - 1)^{n+1}}{2 - \alpha}.$$

Compare the results graphically.

10-2.3. The probability of ruin in the standard gambler's ruin problem is given by

$$u_k = \frac{s^k - s^a}{1 - s^a}, \quad (s \neq 1), \quad u_k = 1 - \frac{k}{a}, \quad (s = 1),$$

where $s = (1 - p)/p$, a is the total stake, k is the gambler's initial stake, and p is the probability that the gambler wins at each play. Design a program to reproduce Figure 2.2, which shows the probability u_k versus k for a given value of a ($a = 20$ in the figure) for a selection of values of p.

10-2.4. The expected duration of the gambler's ruin is given by

$$d_k = \frac{1}{1 - 2p} \left[k - \frac{a(1 - s^k)}{1 - s^a} \right], \quad (s \neq 1), \quad d_k = k(a - k), \quad (s = 1),$$

where $s = (1 - p)/p$, a is the total stake, k is the gambler's initial stake, and p is the probability that the gambler wins at each play. Display the expected duration d_k against k for a given value of a ($a = 20$ is used in the program) for a selection of values of p.

Find the maximum expected duration for different values of p and plot the results.

Chapter 3: Random walks

10-3.1. Design a program to simulate a symmetric random walk which starts at position $k = 0$. Display the walk in a graph with joined successive steps in the k (position of walk) versus n (number of steps) plane as illustrated in Figure 3.3. You could try, for example, 1000 steps.

10-3.2. From the answer to Problem 3.4, the probability that the first return of a symmetric random walk to the origin is given by

$$f_{2n} = (-1)^{n+1} \binom{\frac{1}{2}}{n}.$$

It has been shown that the mean associated with f_n is infinite. Compute

$$\sum_{n=1}^{100} f_{2n},$$

and compare the sum with a simulation of the average length of 200 walks with first returns under 400 steps. (The upper bound of 400 steps is to exclude what might be long-running programs with more than 400 steps to a first return.)

The probability that a first return occurs at the second step is $\frac{1}{2}$. This should be revealed in the output for the means of the lengths of the simulated walks. The output should also show a small number of walks with a large number of steps. As a consequence, several runs should also indicate a high variance.

10-3.3. In Section 3.3 it is shown that the probability that a random walk with parameter p is in state x after n steps from the origin is given by

$$v_{n,x} = \binom{n}{\frac{1}{2}(n+x)} p^{\frac{1}{2}(n+x)} (1-p)^{\frac{1}{2}(n-x)},$$

where n and x are either both odd or both even. In *Mathematica*, define a function of n, x and p to represent $v_{n,x}$. For specific values of n confirm that the expected value of the final position of the walk is $n(2p-1)$. Again for specific values of n (say 16) and p (say $\frac{2}{3}$) show graphically the probability distribution $v_{n,x}$ in terms of x.

10-3.4. Design a program to simulate a two-dimensional symmetric random walk that starts at the origin. Make sure that odd and even values of the numbers of steps are included (see Problems 3.21 and 3.22). Collect the data of the final positions of n trial walks over m steps. Possible numbers are $n = 2000$ and $m = 40$. The data can be displayed in a three-dimensional surface plot.

Let the final positions of the walk be the list (i_k, j_k) for $k = 1, 2, 3, \ldots, n$. Calculate the squares $D_k^2 = i_k^2 + j_k^2$, and find the mean of D_k^2. Try this for different m. What would you guess is the mean value of D_k^2 in terms of m?

Chapter 4: Markov chains

10-4.1. Design a program to find the eigenvalues and eigenvectors of the transition matrix T given by

$$T = \begin{bmatrix} \frac{1}{4} & \frac{1}{8} & \frac{3}{8} & \frac{1}{4} \\ \frac{1}{3} & \frac{1}{6} & \frac{1}{6} & \frac{1}{3} \\ \frac{1}{3} & \frac{1}{3} & 0 & \frac{1}{3} \\ \frac{1}{3} & 0 & 0 & \frac{2}{3} \end{bmatrix},$$

(see also Problem 4.6(c)). If C is the matrix of eigenvectors and D is the diagonal matrix of eigenvalues check by a program that $T = CDC^{-1}$. Find a general formula for T^n. For how many decimal places is $\lim_{t\to\infty} T^n$ correct for T^{10}.

10-4.2. Find the eigenvalues and eigenvectors of the stochastic matrix

$$T = \begin{bmatrix} \frac{1}{4} & \frac{1}{2} & \frac{1}{4} \\ \frac{1}{2} & \frac{1}{4} & \frac{1}{4} \\ \frac{1}{4} & \frac{1}{4} & \frac{1}{2} \end{bmatrix}.$$

Construct a formula for T^n, and find $\lim_{n \to \infty} T^n$.

10-4.3. Devise a program to simulate a two-state Markov chain with transition matrix

$$T = \begin{bmatrix} 1 - \alpha & \alpha \\ \beta & 1 - \beta \end{bmatrix}.$$

Possible values for the parameters are $\alpha = \frac{1}{2}$ and $\beta = \frac{1}{3}$ for, say, 1000 steps. Then count the number of times that states E_1 and E_2 occur in the random process. Compare this output with the theoretical stationary distribution

$$\mathbf{p} = \begin{bmatrix} \frac{\beta}{\alpha + \beta} & \frac{\alpha}{\alpha + \beta} \end{bmatrix}.$$

10-4.4. Devise a program to simulate a three-state (or higher state) Markov chain with transition matrix

$$T = \begin{bmatrix} 1 - \alpha_2 - \alpha_3 & \alpha_2 & \alpha_3 \\ \beta_1 & 1 - \beta_1 - \beta_3 & \beta_3 \\ \gamma_1 & \gamma_2 & 1 - \gamma_1 - \gamma_2 \end{bmatrix}.$$

Test the program with the data from Example 4.3 in which

$$T = \begin{bmatrix} \frac{1}{4} & \frac{1}{2} & \frac{1}{4} \\ \frac{1}{2} & \frac{1}{4} & \frac{1}{4} \\ \frac{1}{4} & \frac{1}{4} & \frac{1}{2} \end{bmatrix}.$$

Count the number of times that states E_1, E_2 and E_3 occur in the random process. Compare this output with the theoretical stationary distribution.

10-4.5. A four-state Markov chain with two absorbing states has the transition matrix

$$S = \begin{bmatrix} 1 & 0 & 0 & 0 \\ \frac{3}{4} & 0 & \frac{1}{4} & 0 \\ 0 & \frac{1}{4} & 0 & \frac{3}{4} \\ 0 & 0 & 0 & 1 \end{bmatrix},$$

(see Problem 4.8(b)). Simulate the chain that starts in state E_2, and determine the proportion of chains that finish in either E_1 or E_2 over, say, 1000 runs. Check this answer to Problem 4.8(b) by finding $\lim_{t \to \infty} T^n$.

10-4.6. As in Problem 4.8(a) a Markov chain has the transition matrix

$$T = \begin{bmatrix} \frac{1}{4} & \frac{1}{4} & \frac{1}{2} \\ 1 & 0 & 0 \\ \frac{1}{2} & \frac{1}{4} & \frac{1}{4} \end{bmatrix}.$$

Confirm the eigenvalues of T and show that T has just two distinct non-trivial eigenvalues. Find the Jordan decomposition matrix J and a matrix C such that $T = CJC^{-1}$. Input a formula for J^n, find T^n and its limit as $n \to \infty$.

10-4.7. This program indicates a method of representing a Markov chain as a *signal flow* of the states $E_1, E_2, \ldots$ against the steps $n = 1, 2, 3, \ldots$. Display the chain for the three-state process with transition matrix

$$T = \begin{bmatrix} \frac{1}{4} & \frac{1}{4} & \frac{1}{2} \\ 1 & 0 & 0 \\ \frac{1}{2} & \frac{1}{4} & \frac{1}{4} \end{bmatrix}$$

(see Problem 4.8(a)).

Apply the signal flow display to the six-state chain

$$T = \begin{bmatrix} \frac{1}{4} & \frac{1}{2} & 0 & 0 & 0 & \frac{1}{4} \\ 0 & 0 & 0 & 0 & 0 & 1 \\ 0 & \frac{1}{4} & 0 & \frac{1}{4} & \frac{1}{2} & 0 \\ 0 & 0 & 0 & 0 & 1 & 0 \\ 0 & 0 & 0 & \frac{1}{2} & \frac{1}{2} & 0 \\ 0 & 0 & 0 & \frac{1}{2} & \frac{1}{2} & 0 \\ 0 & 0 & 1 & 0 & 0 & 0 \end{bmatrix}$$

which appears also in Problem 4.11. The flow diagram should reveal the closed subset.

10-4.8. Form an $n \times n$ matrix whose rows are randomly and independently chosen from the uniform discrete distribution of the numbers $1, 2, 3, \ldots, n$. Transform this into a stochastic matrix T by dividing the elements of each row by the sum of the elements in that row. This is a method of generating a class of *positive* stochastic matrices (see Section 4.7(a)). Find the eigenvalues and eigenvectors of T, T^n and $\lim_{n \to \infty} T^n$.

Chapter 5: Poisson processes

10-5.1. In the Poisson process with parameter λ, the probability that the population is of size n at time t is given by $p_n(t) = (\lambda t)^n e^{-\lambda t}/n!$ (see Section 5.2). Show the first few probabilities in dimensionless form against λt as in Figure 5.1.

Show also that the maximum value of the probability on the graph of $p_n(t)$ lies on the graph of $p_{n-1}(t)$.

10-5.2. Simulate the Geiger counter, which is modelled by a Poisson process with parameter λ. The probability of a recording of a hit in any time interval of duration δt is assumed to be $\lambda \delta t$ with multiple recordings negligible (see Section 5.2). Assuming that the readings start from zero show the growth of the readings against time.

Run the program a number of times over the same time interval, and find the mean value of the data. Compare this value with the theoretical value for the mean of a Poisson process.

Chapter 6: Birth and death processes

10-6.1. Design a program to simulate a simple birth and death process with birth rate λ and death rate μ. Time interval $(0, T)$ should be divided into increments δt, so that the probability that a birth takes place in any interval of duration δt is $\lambda \delta t$, and similarly for a death (see Section 6.5). Show graphically how the population changes with time.

10-6.2. The probability generating function $G(s, t)$ for a simple birth and death process with birth and death rates λ and μ, respectively, is given by

$$G(s, t) = \left[\frac{\mu(1 - s) - (\mu - \lambda s)e^{-(\lambda - \mu)t}}{\lambda(1 - s) - (\mu - \lambda s)e^{-(\lambda - \mu)t}} \right]^{n_0},$$

(see equation (6.23)). Define the function $G(s, t)$ in *Mathematica*, and find the mean and variance of the population size as functions of t.

10-6.3. In a death process, the death-rate is μ and the initial population is n_0. The probability that the population is extinct by time t is given by

$$p_0(t) = (1 - e^{-\mu t})^{n_0}$$

(see equation (6.17)). Show that the mean time T_{n_0} to extinction is given by

$$T_{n_0} = \frac{n_0}{\mu} \sum_{k=0}^{n_0-1} \frac{(-1)^k}{(k + 1)^2} \binom{n_0 - 1}{k},$$

(see Problem 6.29). Write a program to calculate the dimensionless μT_{n_0} from the series. Plot a graph of T_{n_0} versus n_0, say for $n_0 = 1, 2, 3, \ldots, 50$.

Now construct a program to simulate a death process with parameter μ. Calculate a series of extinction times (say, 50?) and compare the mean of these times with the theoretical value from the series.

10-6.4. For a simple birth and death process with equal birth- and death-rates λ, the probability generating function is

$$G(s, t) = \left[\frac{1 + (\lambda t - 1)(1 - s)}{1 + \lambda t (1 - s)} \right]^{n_0},$$

where n_0 is the initial population size (see equation (6.24)). Find the vector function for the probabilities $p_n(t)$ for $n = 0, 1, 2, \ldots, n_1$ as a function of the dimensionless time $\tau = \lambda t$. Show graphically how $p_n(t)$ behaves as τ increases from zero. Possible values are $n_0 = 15$, $n_1 = 30$ and $\tau = 1, 2, 3, 4, 5$.

Chapter 7: Queues

10-7.1. Write a program to simulate a single-server queue with arrivals having a Poisson distribution with parameter λ and with service times having a negative exponential distribution with parameter μ ($\lambda < \mu$). Possible values are $\lambda = 1$, $\mu = 1.1$ with an elapse time of $T = 2000$ in steps of duration $\delta t = 0.1$. By counting the number of times queue lengths $0, 1, 2, \ldots$ occur, a distribution for the simulation can be calculated and represented in a bar chart.

The theoretical probability that r persons are in the queue (including the person being served) in the *stationary* process is given by $p_n = (1 - \rho)\rho^r$ where $\rho = \lambda/\mu$. Compare these probabilities graphically with the bar chart.

10-7.2. Consider the solution for the problem of the queue with r servers given in Section 7.4. The probability p_0 that there is no one queueing is given by equation (7.11):

$$p_0 = 1 \left/ \left[\sum_{n=0}^{r-1} \frac{\rho^n}{n!} + \frac{\rho^r}{(r - \rho)(r - 1)!} \right] \right. .$$

Represent this by a function of the traffic density $\rho_1 = \rho/r$ and the number of servers r. Plot the probabilites against ρ_1 for $r = 1, 2, 3, 4$ servers. The probability p_n that the queue is of length n is given by

$$p_n = \begin{cases} \rho^n p_0/n! & n < r \\ \rho^n p_0/[r^{n-r}r!] & n \geq r \end{cases}$$

Compose a list of the probabilites when $r = 6$ and $n = 0, 1, 2, \ldots, 20$ for the case $\rho_1 = 0.8$. For what value of n does the probability take its largest value?

Display a surface over the grid $r \times n$ where $r = 1, 2, 3, 4, 5, 6$ and $n = 0, 1, 2, \ldots, 20$ again for the case $\rho_1 = 0.8$. For the same parameter, plot the expected queue lengths for $r = 1, 2, \ldots, 20$ excluding those being served.

10-7.3. The fixed service time queue has the probability generating function

$$G(s) = \frac{(1 - \rho)(1 - s)}{1 - s e^{\rho(1-s)}}, \quad 0 < \rho < 1,$$

where $\rho = \lambda/\tau$ and τ is the service time. Using *Mathematica*, check the results in Problem 7.12 for p_0, p_1, p_2 and the expected length of the queue.

10-7.4. In the baulked queue (see Example 7.1) not more than $m \geq 2$ persons are permitted to form a queue. The probability generating function for the queue is (see Problem 7.13) given by

$$G(s) = \frac{(1 - \rho)[1 - (\rho s)^{m+1}]}{(1 - \rho^{m+1})(1 - \rho s)}, \quad \rho \neq 1.$$

Define this function in *Mathematica*, and obtain a list that represents the probability distribution over $i = 0, 1, 2, \ldots, m$.

Now compare these results with a simulated baulked queue. Some possible parameter values are

$$\lambda = 1.2; \quad \mu = 1; \quad \text{incremental time step, } \delta t = 0.1; \, m = 10,$$

run over a time of $t = 2000$.

Chapter 8: Reliability and renewal

10-8.1. In Section 8.2, the reliability function is defined as

$$R(t) = \mathbf{P}(T > t) = 1 - F(t),$$

and the failure rate function is given by $r(t) = f(t)/R(t)$. Obtain the functions $F(t)$, $R(t)$ and $r(t)$ for the gamma density function $f(t) = \lambda^2 t e^{-\lambda t}$. Plot the functions for $\lambda = 1$ over the interval $0 \leq t \leq 5$.

10-8.2. In Problem 8.3 the failure rate function is given by

$$r(t) = \frac{ct}{a^2 + t^2}, \quad t \geq 0, \quad a > 0, \quad c > 1.$$

Find the formula for the reliability function $R(t)$ and the density function $f(t)$. Plot all three functions for $a = 1$ and $c = 2$. What general formula does *Mathematica* give for the mean time to failure in terms of a and c? Find the actual means in the cases (a) $a = 5$, $c = 1.02$, (b) $a = 5$, $c = 2$. Plot the mean time to failure as a surface in terms of a and c for, say, $0.2 < a < 10$ and $1.1 < c < 2$.

10-8.3. Define a reliability function for a process in which n components are in parallel, and each has a time to failure that is independent of all others, and where all times have the same exponential distribution with parameter λ (see Example 8.4). Find the expected mean time to failure for the cases $n = 1, 2, 3, \ldots, 10$, say. How do you expect this mean to behave as $n \to \infty$?

Define a reliability function for the case of three components in parallel exponential distribution but with different parameters λ_1, λ_2 and λ_3.

Chapter 9: Branching and other random processes

10-9.1. A branching process starts with one individual, and the probability that any individual in any generation produces j descendants is given by $p_j = 1/2^{j+1}$ ($j = 0, 1, 2, \ldots$). The probability generating function for the nth generation is given by

$$G_n(s) = \frac{n - (n-1)s}{n + 1 - ns},$$

(see Example 9.1). Find the probability that there are r individuals in the nth generation. Plot the first few terms in the distributions for the cases $n = 1, 2, 3, 4, 5$. Check the mean population size in the nth generation.

10-9.2. The branching process in Problem 9.3 has the probability generating function

$$G(s) = a + bs + (1 - a - b)s^2, \quad a > 0, \quad b > 0, \quad a + b < 1.$$

The process starts with one individual. Using a nesting command find the probabilities that there are $0, 1, 2, \ldots, 31, 32$ individuals in the 5th generation assuming, say, that $a = \frac{1}{4}$ and $b = \frac{1}{3}$. Plot the probabilities against the number of individuals: why do they oscillate? Find also the probability of extinction for general values of a and b.

10-9.3. In Section 9.5, the gambling problem in which a gambler starts with £1 and bets against the house with an even chance of collecting £2 or losing £1 is explained. If the gambler loses then he or she doubles the bet to £2 with the same evens chances. The gambler continues doubling the bet until he or she wins. It was shown that this eventually guarantees that the gambler wins £1. Devise a program that simulates the game, and gives, by taking the mean of a large number of games, an estimate of the expected number of plays until the gambler wins.

10-9.4. Design a program that models the doubling gambling problem in Section 9.5 to simulate the expected value

$$E(Z_{n+1}|Z_0, Z_1, Z_2, \ldots, Z_n) = Z_n,$$

where $Z_0 = 1$. The final outputs should take values in the sequence

$$Z_n = \{-2^n + 2m + 2\}, \qquad (m = 0, 1, 2, \ldots, 2^n - 1).$$

A possible case to try is $n = 5$ over 5000 trials: the results can be expressed graphically in a bar chart.

10-9.5. In Section 9.8 on an iterative scheme for the simple epidemic, the probability $q_n(\tau)$ that there are n susceptibles in the epidemic population at (non-dimensional) time $\tau (= \beta t)$ satisfies the differential-difference equations

$$\frac{dq_n(\tau)}{d\tau} = (n+1)(n_0-n)q_{n+1}(\tau) - n(n_0+1-n)q_n(\tau), \quad (0 \leq n \leq n_0-1),$$

and

$$\frac{dq_{n_0}(\tau)}{d\tau} = -n_0 q_{n_0}(\tau).$$

Construct a simple program to solve the equations iteratively for the first few probabilities $q_{n_0}, q_{n_0-1}, \ldots$. Compare the results with those of Example 9.9 in which $n_0 = 4$.

Appendix

Abbreviations

iid — independent and identically distributed.
mgf —moment generating function.
pgf — probability generating function.
rv — random variable.

Notation

A, B — sets or events.
$A \cup B$ — union of A and B.
$A \cap B$ — intersection of A and B.
A^c — complement of A.
N, X, Y — random variables.
$\mathbf{P}(X = x)$ — probability that the random variable $X = x$.
$\mathbf{E}(X)$ — mean or expected value of the random variable X.
$\mathbf{V}(X)$ — variance of the random variable X.
μ — alternative symbol for mean or expected value.
σ^2 — alternative symbol for variance.
Factorial function, $n! = 1 \cdot 2 \cdot 3 \cdots n$; $0! = 1$.
Gamma function, $\Gamma(n) = (n - 1)\Gamma(n - 1) = (n - 1)!$.
Binomial coefficients:

$$\binom{n}{r} = \frac{n!}{(n - r)!r!}, \qquad n, r \text{ positive integers } n \geq r;$$

$$\binom{a}{r} = \frac{a(a - 1) \cdots (a - r + 1)}{1 \cdot 2 \cdot 3 \cdots r}, \qquad a \text{ any real number.}$$

Power series

Exponential function:

$$e^{-t} = \sum_{n=0}^{\infty} \frac{t^n}{n!} = 1 + t + \frac{t^2}{2!} + \frac{t^3}{3!} + \cdots \quad \text{(for all } t\text{)};$$

Binomial expansions:

$$(1+s)^n = \sum_{r=0}^{n} \binom{n}{r} s^r$$

$$= 1 + ns + \frac{n(n-1)}{2!} s^2 + \cdots + ns^{n-1} + s^n, \quad (n \text{ is a positive integer})$$

$$(1+s)^a = \sum_{r=0}^{\infty} \binom{a}{r} s^r$$

$$= 1 + as + \frac{a(a-1)}{2!} s^2 + \frac{a(a-1)(a-2)}{3!} s^3 + \cdots$$

$$(|s| < 1, a \text{ any real number}).$$

Probability generating function:

$$G(s, t) = \sum_{n=0}^{\infty} p_n(t) s^n,$$

$$G(1, t) = 1, \quad \text{mean } = \mu = G_s(1, t),$$

$$\text{variance } = \sigma^2 = G_{ss}(1, t) + G_s(1, t) - G_s(1, t)^2.$$

Integrals

$$\int_0^{\tau} e^{-\lambda t} \, dt = \frac{1}{\lambda}(1 - e^{-\lambda \tau}), \qquad \int_0^{\infty} e^{-\lambda t} \, dt = \frac{1}{\lambda}.$$

$$\int_0^{\tau} t e^{-\lambda t} \, dt = \frac{1}{\lambda^2}[1 - (1 + \lambda \tau)e^{-\lambda \tau}], \qquad \int_0^{\infty} t e^{-\lambda t} \, dt = \frac{1}{\lambda^2}.$$

$$\int_0^{\infty} t^n e^{\lambda t} \, dt = \frac{n!}{\lambda^{n+1}}.$$

$$\int_0^{\infty} e^{-\lambda t^2} \, dt = \frac{1}{2}\sqrt{\frac{\pi}{\lambda}}.$$

Matrix algebra

Transition matrix ($m \times m$):

$$T = [p_{ij}] = \begin{bmatrix} p_{11} & p_{12} & \cdots & p_{1m} \\ p_{21} & p_{22} & \cdots & p_{2m} \\ \vdots & \vdots & \ddots & \vdots \\ p_{m1} & p_{m2} & \cdots & p_{mm} \end{bmatrix},$$

row-stochastic:

$$\sum_{j=1}^{m} p_{ij} = 1, \quad p_{ij} \geq 0.$$

Matrix product of $A = [a_{ij}]$, $B = [b_{ij}]$, both $m \times m$:

$$AB = [a_{ij}][b_{ij}] = \left[\sum_{k=1}^{m} a_{ik} b_{kj} \right].$$

Diagonal matrix ($m \times m$) with diagonal elements $\lambda_1, \lambda_2, \ldots, \lambda_m$:

$$D = \begin{bmatrix} \lambda_1 & 0 & \cdots & 0 \\ 0 & \lambda_2 & \cdots & 0 \\ \vdots & \vdots & \ddots & \vdots \\ 0 & 0 & \cdots & \lambda_m \end{bmatrix}.$$

Identity matrix:

$$I_m = \begin{bmatrix} 1 & 0 & \cdots & 0 \\ 0 & 1 & \cdots & 0 \\ \vdots & \vdots & \ddots & \vdots \\ 0 & 0 & \cdots & 1 \end{bmatrix}.$$

Characteristic equation for the eigenvalues $\lambda_1, \lambda_2, \ldots, \lambda_m$ of T:

$$|T - \lambda I_m| = 0.$$

Eigenvectors $\mathbf{r}_1, \mathbf{r}_2, \ldots, \mathbf{r}_m$ satisfy:

$$[T - \lambda_i I_m] \mathbf{r}_i = \mathbf{0}, \quad (i = 1, 2, \ldots, m).$$

Probability distributions

Discrete distributions:

Distribution	$P(X = x)$	Mean	Variance
Bernoulli	$p^x(1-p)^{1-x}$ $(x = 0, 1)$	p	$p(1-p)$
Binomial	$\binom{n}{x}p^x(1-p)^{n-x}$ $(x = 0, 1, 2, \ldots, n)$	np	$np(1-p)$
Geometric	$(1-p)^{x-1}p$ $(x = 1, 2, \ldots)$	$1/p$	$(1-p)/p^2$
Negative binomial	$\binom{x-1}{r-1}p^x(1-p)^{x-r}$ $(x = r, r+1, \ldots)$	r/p	$r(1-p)/p^2$
Poisson	$e^{-\alpha}\alpha^x/x!,$ $(x = 0, 1, 2, \ldots)$	α	α
Uniform	$1/n,$ $(x = r, r+1, \ldots, r+n-1)$	$\frac{1}{2}(n+1)$	$\frac{1}{12}(n^2-1)$

Continuous distributions:

Distribution	Density, $f(x)$	Mean	Variance
Exponential	$\alpha e^{-\alpha x}$ $(x \geq 0)$	$1/\alpha$	$1/\alpha^2$
Normal, $[N(\mu, \sigma^2)]$	$\dfrac{1}{\sigma\sqrt{2\pi}}\exp\left[-\dfrac{(x-\mu)^2}{2\sigma^2}\right]$ $-\infty < x < \infty$	μ	σ^2
Gamma	$\dfrac{\alpha^n}{\Gamma(n)}x^{n-1}e^{-\alpha x},$ $(x \geq 0)$	n/α	n/α^2
Uniform	$\begin{cases} 1/(b-a) & a \leq x \leq b \\ 0 & \text{for all other } x \end{cases}$	$\frac{1}{2}(a+b)$	$\frac{1}{12}(b-a)^2$
Weibull	$\alpha\beta x^{\beta-1}e^{-\alpha x^\beta},$ $(x \geq 0)$	$\alpha^{-1/\beta}\Gamma(\beta^{-1}+1)$	$\alpha^{-2/\beta}[\Gamma(2\beta^{-1}+1)$ $-\{\Gamma(\beta^{-1}+1)\}^2]$

References and further reading

Abell, M. L., Braselton, J. P. and Rafter, J. A. (1998) *Statistics with Mathematica*, Academic Press, New York.

Bailey, N. T. J. (1964) *The Elements of Stochastic Processes*, 6th edn, John Wiley, New York.

Feller, W. (1968) *An Introduction to Probability Theory and its Applications*, 3rd edn, John Wiley, New York.

Grimmett, G. R. and Stirzaker, D. R. (1982) *Probability and Random Processes*, Clarendon Press, Oxford.

Grimmett, G. R. and Welsh, D. (1986) *Probability: an Introduction*, Clarendon Press, Oxford.

Hogg, R. V. and Tabis, E. A. (2001) *Probability and Statistical Inference*, Prentice-Hall, New Jersey.

Jones, P. W. and Smith, P. (1987) *Topics in Applied Probability*, Keele Mathematical Education Publications, Keele University.

Jordan, D. W. and Smith, P. (1997) *Mathematical Techniques*, 2nd edn, Oxford University Press.

Lawler, G. F. (1995) *Introduction to Stochastic Processes*, Chapman & Hall, New York.

Luenberger, D. G. (1979) *Introduction to Dynamic Systems*, John Wiley, New York.

McColl, J. H. (1995) *Probability*, Arnold, London.

Meyer, P. L. (1970) *Introductory Probability and Statistical Applications*, 2nd edn, Addison-Wesley, Reading, MA.

Open University (1988) *Applications of Probability*, Course Units for M343, Open University Press, Milton Keynes.

Ross, S. (1976) *A First Course in Probability*, Macmillan, New York.

Tuckwell, H. C. (1988) *Elementary Applications of Probability Theory*, Chapman & Hall, London.

Tuckwell, H. C. (1989) *Stochastic Processes in the Neurosciences*, SIAM Publications, Philadelphia.

Wolfram, S. (1996): *The Mathematica Book*, 3rd edn, Cambridge University Press.

Answers and comments on some of the end-of-chapter problems

Chapter 1

1.2. (c) $(A \cap B) \cap C^c$; (e) $[A \cap (B \cup C)^c] \cup [B \cap (A \cup C)^c] \cup [C \cap (A \cup B)^c]$.

1.3. (a) 0.6; (c) 0.4.

1.6. The mean value is $\frac{3}{4}$.

1.7. The mean is α and the variance is also α.

1.8. The probability generating function is $ps/(1 - qs)$. The mean is $1/p$ and the variance is q/p^2.

1.10. (a) 11/36; (b) 5/18; (c) 5/36; (d) 1/18; (e) 7/36.

1.11. The expected value of the face values is 7, and the variance is 35/6.

1.12. There are 216 possible outcomes, the probability that two dice have the same face values and the third is different is 5/12.

1.13. 17/52.

1.14. Cumulative distribution function is $F(x) = 1 - \frac{1}{2} e^{-(x-a)/a}$.

1.15. $G(s) = ps/[1 - s(1 - p)]$.

1.16. The expected value of the volume of the tetrahedron is $\mu/[12(3\sigma^2 + \mu^2)]$.

1.17. The expected value is given by

$$\mathbf{E}(X^p) = \frac{n!}{\alpha^p (n + p)!}.$$

1.18. The probabilities are

$$p_0 = \frac{1 - \alpha}{1 + \alpha}, \quad p_n = \frac{2\alpha^n}{(1 + \alpha)^{n+1}}.$$

1.19. The moment generating function is

$$M_X(t) = \frac{1}{b - a} \sum_{n=1}^{\infty} \left(\frac{b^n - a^n}{n!} \right) t^{n-1},$$

and

$$\mathbf{E}(X^n) = \frac{b^{n+1} - a^{n+1}s}{(n + 1)(b - a)}.$$

1.21. (a) μ, $p(1 - p)$; (b) $p/[1 - (1 - p)s]^2$, $(1 - p)/p^2$; (c) $r(r + 1)(1 - p)^r p^2/[(1 - ps)^{r+2}]$, $pr/(1 - p)^2$.

1.22. Use the conditional formula

$$f_{N|M} = \frac{f_{M|N} f_N}{f_M},$$

so that the conditional expectation

$$\mathbf{E}(N|M) = M + \lambda(1 - p).$$

1.23. $(0.95)^5 \approx 0.774$ and $0.9975^5 \approx 0.9876$.

1.24. The mean of the random variable is

$$\mu = -\frac{\alpha}{(1 - \alpha)\ln(1 - \alpha)}.$$

Chapter 2

2.1. (a) Probability of ruin is $u_k \approx 0.132$, and the expected duration is $d_k \approx 409$; (c) $u_k = 0.2$, $d_k = 400$.

2.2. The probability $p = 0.498999$ to 6 digits: note that the result is very close to $\frac{1}{2}$.

2.3. (a) $u_k = A + 3^k B$; (c) $u_k = A + (1 + \sqrt{2})^k B + (1 - \sqrt{2})^k$.

2.4. (a) $u_k = (625 - 5^k)/624$; (c) $d_k = k(10 - k)$.

2.6. The expected duration is $k(a - k)$.

2.8. The game would be expected to take four times as long.

2.9. The probabilities of the gambler winning in the two cases are 2.23×10^{-4} and 2.06×10^{-6}.

2.10. $p_n = \frac{2}{3} + \frac{1}{3}(-\frac{1}{2})^n$, $n = 1, 2, 3, \ldots$.

2.11. $p_n = [1 - (\alpha - 1)^{n+1}]/(2 - \alpha)$.

2.14. Game extended by $d_{2k,2a} - d_{k,a}$ where

$$d_{m,n} = \frac{1}{1 - 2p}\left[m - \frac{a(1 - s^m)}{(1 - s^n)}\right].$$

2.18. $u_k = k/100$.

2.20. The expected duration of the game is about 99 plays.

2.22. For $s < 1$ and large a,

$$k \approx \frac{\ln 2}{\ln s} + s^a.$$

2.23. (a) $\alpha_k = \frac{1}{2}$; (c) $\alpha_k = \sin[(k - \frac{1}{2})\pi a]/[2\cos(\pi/2a)\sin(k\pi/a)]$.

2.25. (c) $[(n - 1)/n]^{n-1}$.

Chapter 3

3.2. The probability that the walker is at the origin at step 8 is

$$P(X_8 = 0) = \frac{1}{2^8}\binom{8}{4} \approx 0.273.$$

The probability that it is not the first visit there is approximately 0.234.

3.4. (a)

$$f_n = \begin{cases} (-1)^{n+1}\binom{\frac{1}{2}}{n} & (n \text{ even}) \\ 0 & (n \text{ odd}) \end{cases}.$$

3.5. Treat the problem as a symmetric random walk with a return to the origin: the probability is

$$v_{n,0} = \begin{cases} \dfrac{1}{2^n}\binom{n}{\frac{1}{2}n} & (n \text{ even}) \\ 0 & (n \text{ odd}) \end{cases}.$$

3.7. The mean number of steps to the first return is $4pq/|p - q|$.

3.14. The probability that the walk ever visits $x > 0$ is $(1 - |p - q|)/(2q)$.

3.17. (b) The probability that the walk is at O after n steps is 0 if n and N are neither both even nor odd, and

$$\frac{1}{2^n}\left(\binom{n}{\frac{1}{2}n} + \binom{n}{\frac{1}{2}(n+N)} + \binom{n}{\frac{1}{2}(n-N)}\right)$$

if n and N are both even, and

$$\frac{1}{2^n}\left(\binom{n}{\frac{1}{2}(n+N)} + \binom{n}{\frac{1}{2}(n-N)}\right)$$

if n and N are both odd.

3.18. (a) 0.217; (b) 0.415.

3.19. $p = (1 + \sqrt{5})/4$.

3.20. The probability that they are both at the origin at step n is 0 if n is odd, and

$$\frac{1}{2^{2n}}\binom{n}{\frac{1}{2}n}^2.$$

3.21. $p_{2n} \sim 1/(n\pi)$, ignoring the $\frac{1}{4}$ for large n.

Chapter 4

4.1. $p_{23} = 5/12$; $\mathbf{p}^{(1)} = [\frac{173}{720}, \frac{1}{3}, \frac{307}{720}]$.

4.2. $p_{31}^{(2)} = \frac{13}{36}$.

4.3. $\mathbf{p}^{(3)} = [943, 2513]/3456$; eigenvalues of T are $1, \frac{1}{12}$;

$$T^n \to \frac{1}{11} \begin{bmatrix} 3 & 8 \\ 3 & 8 \end{bmatrix} \quad \text{as} \quad n \to \infty.$$

4.5. Eigenvalues are

$$a + b + c, \quad \frac{1}{2}[2a - b - c \pm i\sqrt{3}|b - c|].$$

(a) The eigenvalues are $1, \frac{1}{4}, \frac{1}{4}$ with corresponding eigenvectors

$$\begin{bmatrix} 1 \\ 1 \\ 1 \end{bmatrix}, \quad \begin{bmatrix} -1 \\ 1 \\ 0 \end{bmatrix}, \quad \begin{bmatrix} -1 \\ 0 \\ 1 \end{bmatrix}.$$

4.6. (a)

$$T^n = \frac{1}{11} \begin{bmatrix} 4 + 7(-\frac{3}{8})^n & 7 - 7(-\frac{3}{8})^n \\ 4 - (-\frac{3}{8})^n & 7 + (-\frac{3}{8})^n \end{bmatrix}.$$

(b)

$$\lim_{n \to \infty} T^n = \frac{1}{91} \begin{bmatrix} 28 & 9 & 12 & 42 \\ 28 & 9 & 12 & 42 \\ 28 & 9 & 12 & 42 \\ 28 & 9 & 12 & 42 \end{bmatrix}.$$

4.7. 40% of the days are sunny.

4.8. (c) The eigenvectors are $\frac{1}{3}, \frac{1}{3}$ and 1. The invariant distribution is $[\frac{1}{4}, 0, \frac{3}{4}]$.

4.9. (a) Eigenvalues are $-\frac{1}{8}, -\frac{1}{8}, 1$.

(b) A matrix C of transposed eigenvectors is given by

$$C = \begin{bmatrix} 1 & \frac{5}{12} & 1 \\ -4 & -\frac{5}{2} & 1 \\ 1 & 1 & 1 \end{bmatrix}.$$

(c) The invariant distribution is $[28, 15, 39]/82$.

(d) The eigenvectors are $-\frac{1}{4}, \frac{1}{12}, 1$.

(e) The limiting matrix is

$$\lim_{n \to \infty} T^n = \begin{bmatrix} 1 & 0 & 0 & 0 \\ \frac{2}{3} & 0 & \frac{1}{3} & 0 \\ 0 & 0 & 1 & 0 \\ \frac{1}{3} & 0 & \frac{2}{3} & 0 \end{bmatrix}.$$

4.10. $f_1 = 1$; $f_2 = 1$; $f_3 = \frac{1}{12}$; $f_4 = \frac{1}{12}$. Every row of $\lim_{n \to \infty} T^n$ is $\{\frac{2}{3} \quad \frac{1}{3} \quad 0 \quad 0\}$.

4.11. E_4 and E_5 form a closed subset with a stationary distribution $[\frac{1}{3} \quad \frac{2}{3}]$.

4.12. All states except E_7 have period 3.

4.15. $\mu_1 = 1 + a + ab + abc$. E_1 is an ergodic state.

4.17. Probability of an item being completed is $(1 - p - q)^2/(1 - q)^2$.

4.18. Probability of an item being completed with N stages is $(1 - p - q)^N/(1 - q)^N$.

4.19. $\mu_3 = 2$.

4.21. The mean of the returns to E_1 is 6.

Chapter 5

5.3. The mean and variance are both λt.

5.6. The mean number of calls received by time t is

$$\mu(t) = \int_0^t \lambda(s)\, ds.$$

5.7. The mean number of calls received by time t is $at + (b/\omega)\sin(\omega t)$.

5.8. The mean reading at time t is $n_0 + \lambda t$.

5.9. (a) $\mathbf{P}[N(3) = 6] = 0.00353$; (c) $\mathbf{P}[N(3.7) = 4 | N(2.1) = 2] = 0.144$; (e) $\mathbf{P}[N(7) - N(3) = 3] = 0.180$.

5.10. The bank should employ 17 operators.

5.11. The expected value of the switch-off times is n/λ.

Chapter 6

6.1. The probability that the original cell has not divided at time t is $e^{-\lambda t}$. The variance at time t is $e^{2\lambda t} - e^{\lambda t}$.

6.3. The probability that the population has halved at time t is

$$p_{\frac{1}{2}n_0}(t) = \binom{n_0}{\frac{1}{2}n_0} [e^{-\mu t}(1 - e^{-\mu t})]^{\frac{1}{2}n_0}.$$

The required time is

$$-\frac{1}{\mu} \ln\left(\frac{\mu}{1 + \mu}\right).$$

6.4. (b) $p_0(t) = 0$, $\quad p_n(t) = e^{-\lambda t}(1 - e^{-\lambda t})^{n-1}$, $\quad (n = 1, 2, 3, \ldots)$.

6.5.
$$p_n(t) = \binom{r}{n}\left(\frac{2}{2+t}\right)^r \left(\frac{t}{2}\right)^n.$$

The mean is $rt/(2 + t)$.

6.6 (b), (c) The probability of extinction at time t is

$$G(0, t) = \left[\frac{\mu - \mu e^{-(\lambda - \mu)t}}{\lambda - \mu e^{-(\lambda - \mu)t}} \right]^{n_0} \rightarrow \begin{cases} (\mu/\lambda)^{n_0} & \text{if } \lambda > \mu \\ 1 & \text{if } \lambda < \mu. \end{cases}$$

(d) The variance is

$$n_0(\lambda + \mu)e^{(\lambda - \mu)t}[e^{(\lambda - \mu)t} - 1]/(\lambda - \mu).$$

6.7. The probability of ultimate extinction is $e^{-\lambda/\mu}$.

6.19. The maximum value of $p_n(t)$ is

$$\binom{n-1}{n_0-1} n_0^{n_0} n^{-n}(n-n_0)^{n-n_0}.$$

6.21. The mean population size is

$$n_0 \exp\left[-\int_0^t \mu(s)\, ds \right],$$

and

$$\mu(t) = \frac{\alpha}{1 + \alpha t}.$$

6.22. Ultimate extinction is certain.

6.23. The probability that the population size is n at time t is given by

$$p_n(t) = \frac{2\mu^n e^{-nt}}{(1 + \mu e^{-t})^{n+1}},$$

and its mean is $2\mu e^{-t}$.

6.24. The probability that the number of descendants becomes zero is μ/λ.

6.26.

$$p_{0,j} = \begin{cases} 0, & j > 0 \\ 1, & j = 0 \end{cases}$$

6.28. The result

$$\int_0^c \frac{u e^{-u}}{(1 - a e^{-u})^2}\, du = -\frac{c}{a(1 - a e^{-u})} + \frac{1}{a} \ln\left[\frac{e^c - a}{1 - a} \right]$$

can be obtained by integration by parts. Then let $c \to \infty$.

6.31. The probability generating function is

$$G(s, t) = \frac{1 - e^{-\mu t} - (1 - e^{-\lambda t} - e^{-\mu t})s}{1 - (1 - e^{-\lambda t})s}.$$

6.32.

$$\frac{a\lambda}{\lambda + \mu} + \left[a_0 - \frac{a\lambda}{\lambda + \mu} \right] e^{-(\lambda + \mu)t}.$$

Chapter 7

7.1. (a) $p_n = 1/2^{n+1}$; (c) $7/8$.

7.3.
$$p_0 = 1/\left[1 + \sum_{n=1}^{\infty} \frac{\lambda_0 - \lambda_{n-1}}{\mu_1 - \mu_n}\right].$$

The expected length of the queue is $1/\mu$.

7.4. If $m = 3$ and $\rho = 1$, then the expected length is $\frac{3}{2}$.

7.5. The mean and variance are respectively $\lambda/(\mu - \lambda)$ and $\lambda(2\lambda - \mu)/(\mu - \lambda)^2$.

7.6. $\rho \approx 0.74$.

7.8. 109 telephones should be manned.

7.9. If the expected lengths are $\mathbf{E}_M(N)$ and $\mathbf{E}_D(N)$ respectively, then $\mathbf{E}_D(N) < \mathbf{E}_M(N)$ assuming $0 < \rho < 1$, where $\rho = \lambda/\mu$.

7.10. $0 \le \rho \le \sqrt{5} - 1$.

7.11. The required probability generating function is
$$G(s) = \left(\frac{2 - \rho}{2 + \rho}\right)\left(\frac{2 + \rho s}{2 - \rho s}\right).$$

7.12. (b) The expected length of the queue is $\rho(1 - \frac{1}{2}\rho)/(1 - \rho)$ and its variance is
$$\frac{\rho}{12(1 - \rho)^2}(12 - 18\rho + 10\rho^2 - \rho^3).$$
(c) $\rho = 3 - \sqrt{5}$.

7.13. The expected length is $\frac{1}{2}n$.

7.14. The expected length is $\rho^2/(1 - \rho)$.

7.15. The mean service time is $7/4$ minutes.

7.17. The service has a uniform density function given by
$$f(t) = \begin{cases} 1/(\tau_2 - \tau_1) & \tau_1 \le t \le \tau_2 \\ 0 & \text{elsewhere} \end{cases}$$

The expected length of the queue is $9/4$.

7.19. The expected waiting time is $\frac{1}{4}\mu$.

7.20. The length of the waiting list becomes $0.89 + O(\varepsilon)$.

7.21. The expected value of the waiting time is
$$\frac{1}{\mu}\left(1 + \frac{p_0 \rho^r}{(r - 1)!(r - \rho)^2}\right),$$
where p_0 is given by (7.11).

7.22. The probability that no one is waiting except those being served is
$$1 - \frac{p_0 \rho^n}{(r - 1)!(r - \rho)},$$
where p_0 is given by (7.11).

7.23. For all booths the expected number of cars queueing is

$$\frac{(r-1)\lambda}{\mu(r-1) - \lambda},$$

and the number of extra vehicles queueing is

$$\frac{\lambda^2}{[\mu(r-1) - \lambda][\mu r - \lambda]}.$$

7.24. $\mu = 6\lambda$.

7.27. A baulked queue of length r with r servers.

Chapter 8

8.1. The reliability function is

$$R(t) = \begin{cases} 1 & t \le 0 \\ (t_1 - t)/(t_1 - t_0) & 0 < t < t_1 \\ 0 & t \ge t_1 \end{cases}$$

The expected life of the component is $\frac{1}{2}(t_0 + t_1)$.

8.2. The failure rate function is $r(t) = \lambda^2 t/(1 + \lambda t)$. The mean and variance of the time to failure are respectively $2/\lambda$ and $2/\lambda^2$.

8.3. The probability density function is given by

$$f(t) = \frac{ca^c t}{(a^2 + t^2)^{(c+2)/2}}.$$

8.4. The reliability function is

$$R(t) = \begin{cases} e^{-\lambda_1 t^2}, & 0 < t < t_0 \\ e^{-[(\lambda_1 - \lambda_2)t_0(2t - t_0) + \lambda_2 t^2]}, & t > t_0 \end{cases}$$

8.5. The expected time to maintenance is 61.2 hours.

8.6. The expected time to failure is n/λ.

8.7. The mean time to failure of the generator is $1/\lambda_f$ and the mean time for the generator to be operational again is $(1/\lambda_f) + (1/\lambda_r)$. The probability that, at time τ, measured from the breakdown of the main generator, the stand-by generator fails before the main one is repaired is

$$e^{-\lambda_r \tau}(1 - e^{-\lambda_s \tau}).$$

8.8. The relation between the parameters is given by

$$\lambda = \frac{1}{1000} \ln \left[\frac{0.999 - (1 + 1000\mu)e^{-1000\mu}}{1 - (1 + 1000\mu)e^{-1000\mu}} \right].$$

8.9. Three spares should be carried.

8.10. (a) $\mathbf{P}(T_1 < T_2) = \lambda_1/(\lambda_1 + \lambda_2)$.

(b) $\mathbf{P}(T_1 < T_2) = \lambda_1^2(\lambda_1 + 3\lambda_2)/(\lambda_1 + \lambda_2)^3$.

8.11. $r(t) \to \lambda_1$ if $\lambda_2 > \lambda_1$, and $r(t) \to \lambda_2$ if $\lambda_2 < \lambda_1$.

8.12. $\mathbf{P}(S_3 \le 3) = 1 - (1 + \lambda t + \frac{1}{2}t^2\lambda^2)e^{-\lambda t}$.

8.13. $\mathbf{P}(S_2 \le 2) = \frac{1}{2}(t/k)^2$, $(0 < t < k)$, $\mathbf{P}(S_2 \le 2) = t(2k - t)/(2k^2)$, $(t > k)$.

Chapter 9

9.1. $p_{n,j} = [(2^n - 1)/2^n]^{j-1}$. The mean population size of the nth generation is 2^n.

9.2. The probability generating function of the second generation is

$$G_2(s) = \frac{(1 - p)(1 - ps)}{(1 - p + p^2) - ps}.$$

The mean population size of the nth generation is $\mu_n = [p/(1 - p)]^n$

9.3. The maximum population of the nth generation is 2^n. It is helpful to draw a graph in the (a, b) plane to show the region where extiction is certain.

9.4. The variance of the nth generation is

$$\frac{\lambda^{n+1}(\lambda^n - 1)}{(\lambda - 1)} = \lambda^{n+1}(\lambda^{n-1} + \lambda^{n-2} + \cdots + 1).$$

9.5. The expected size of the nth generation is

$$\mu_n = \lambda^n \tanh^n \lambda.$$

9.7. (a) The probability that the population size of the nth generation is m is

$$p_n = \frac{n^{m-2}}{(n + 1)^{m+2}}(m + 2n^2 - 1).$$

(b) The probability of extinction by the nth generation is $[n/(n + 1)]^2$.

9.8. $\mu_n = r\mu^n$.

9.9. (b)

$$\mathbf{E}(Z_n) = \begin{cases} (1 - \mu^{n+1})/(1 - \mu) & \mu \neq 1 \\ n + 1 & \mu = 1 \end{cases}$$

(d) The variance of Z_n is

$$\mathbf{V}(Z_n) = \frac{\sigma^2(1 - \mu^n)(1 - \mu^{n+1})}{(1 - \mu)(1 - \mu^2)},$$

if $\mu \neq 1$, or $\frac{1}{2}\sigma^2 n(n + 1)$ if $\mu = 1$.

9.10. (a) $\mu_n = \mu^n$ and $\sigma_n^2 = \mu^n(\mu^n - 1)/(\mu - 1)$ if $\mu \neq 1$.

   (b) $\mu_n = [p/(1 - p)]^n$.

   (c)

$$\mu_n = \left[\frac{rp}{1 - p}\right]^n, \quad \sigma_n^2 = \frac{(pr)^n}{(1 - p)^{n+1}(rp - 1 + p)}\left[\left(\frac{rp}{1 - p}\right)^n - 1\right],$$

   if $\mu \neq 1$.

9.11. The probability of extinction is $[2 - p - \sqrt{(4p - 3p^2)}]/(2p)$.

9.12. (a) $1/[n(n + 1)]$.

   (b) $1/n$.

   The mean number of generations to extinction is infinite.

9.13. The mean number of plants in generation n is $(p\lambda)^n$.

9.14. The probability of extinction by the nth generation is

$$G_n(0) = \frac{(1 - p)[p^n - (1 - p)^n]}{p^{n+1} - (1 - p)^{n+1}}.$$

9.21. $V(Z_n) = 2^n - 1$ and $V[E(Z_n|Z_0, Z_1, \ldots, Z_{n-1})] = V(Z_{n-1}) = 2^{n-1} - 1$.

9.22. The expected position of the walk after 20 steps is $10(p_1 + p_2 - q_1 - q_2)$.

9.26. $G_2(s) = n_0 s^{n_0 - 1}(s - 1)[(2s - 1)n_0 - 2]/2!$.

Index